航空英语系列教材

航空英语 读写教程

总主编：徐 翰
主　编：徐筱秋
副主编：舒宏伟　杨 梅　胡 妮
编　者：胡爱国　齐昂昆　刘爱玲

U0150770

外语教学与研究出版社
北京

图书在版编目（CIP）数据

航空英语读写教程 / 徐筱秋主编；胡爱国，齐昂昆，刘爱玲编 . —— 北京：外语教学
与研究出版社，2022.1
航空英语系列教材 / 徐翰总主编
ISBN 978-7-5213-3289-6

Ⅰ . ①航… Ⅱ . ①徐… ②胡… ③齐… ④刘… Ⅲ . ①航空－英语－阅读教学－高等
学校－教材②航空－英语－写作－高等学校－教材 Ⅳ . ①V2

中国版本图书馆 CIP 数据核字 (2022) 第 005125 号

出 版 人　王　芳
项目策划　陈丽丽
责任编辑　陈丽丽
责任校对　袁敬娜
装帧设计　付玉梅
出版发行　外语教学与研究出版社
社　　址　北京市西三环北路 19 号（100089）
网　　址　http://www.fltrp.com
印　　刷　北京市大天乐投资管理有限公司
开　　本　787×1092　1/16
印　　张　10.5
版　　次　2022 年 3 月第 1 版　2022 年 3 月第 1 次印刷
书　　号　ISBN 978-7-5213-3289-6
定　　价　39.90 元

购书咨询：（010）88819926　电子邮箱：club@fltrp.com
外研书店：https://waiyants.tmall.com
凡印刷、装订质量问题，请联系我社印制部
联系电话：（010）61207896　电子邮箱：zhijian@fltrp.com
凡侵权、盗版书籍线索，请联系我社法律事务部
举报电话：（010）88817519　电子邮箱：banquan@fltrp.com
物料号：332890001

前　言

2021 年是中国共产党成立 100 周年，也是新中国航空事业发展 70 周年。航空科技是 20 世纪以来发展最为迅速、对人类生产生活影响最大的科技领域之一。一个国家在航空领域的重要成果，充分彰显着该国科学技术的先进水平。

随着我国航空事业的不断发展，国家和社会对专业人才的外语水平要求不断提升。培养全面掌握航空英语专业知识的人才，有助于了解全球航空领域动态信息，推动世界科技创新发展，促进全球航空科技交流与合作。

江西是新中国航空工业的摇篮，也是全国重要的航空产业研发和生产基地。近年来，江西已把航空产业列为全省一号战略性新兴产业，并出台了一系列政策措施，着力打造航空强省。在这个背景之下，客观上对高层次、应用型的航空类专业人才的需求日益迫切。与此同时，对于省内高校而言，围绕航空企业对专业人才与外语水平的素质要求，立足市场需求，定位人才培养目标，规划高校教材，努力为航空企业量身打造人才，也有着十分重要的战略意义。

本教材是依据《大学英语教学指南》（2020 年版），基于 OBE 教学理念而编写的航空类英语读写教材。该教材有助于学生了解当前航空技术的发展趋势，熟悉航空专业术语与词汇，掌握专门用途英语的特定表达方式，提高科技类文体翻译能力和写作表达能力，并做好航空语言和相关基础知识的储备。本教材的出版目的是服务国家战略发展，立足航空企业与社会需求，满足 ESP 教学需要，切实提高专业学生的创业与就业能力。

本教材在选材上注重新颖性、前瞻性和科学性。阅读素材主要选自学术期刊、学术网站、专业著作、科普书籍和英文报纸等资源。课文内容既典型又新颖，充分体现了航空学科特点。教材主题丰富，包括中国航天、世界航天、民航飞行安全、民航女飞行员、航天健康风险、太空奥秘、民航环保飞行、航空与娱乐等八个单元。每个单元包含两篇与单元主题相关的阅读文章。其中，Intensive reading 的练习部分，设计了常见科技阅读词汇练习、中外航空科技发展方面的段落翻译，以及篇章填空练习、创意写作练习、批判性思维练习等题型。该部分旨在加深学生对各单元主题的理解与思考，提高语言的实践应用能力。Further reading 的练习设计，主要以文本理解为目标。每个单元后均配有学术写作相关模块训练，如摘要写作、撰写实验报告等等。本教材既可以作为相关专业大学生的 ESP 英语读写教材，也可以作为航空类专业大二、大三学生的专业英语读写教材。

本教材是南昌航空大学"校本特色教材"的重点建设项目之一。教材的编写与出版得到了学校各级部门和外国语学院全体同仁的大力支持。另外，钟立胜主任在教材编写过程中给予了专业知识指导，外教 Olan Patrick Rynn 对教材的英语表达进行了审核，研究生程雅、王菲、王一新参与了教材的编写工作。在此表示衷心感谢！

由于编者水平有限，疏漏和不妥之处在所难免，恳请读者不吝指正！

总主编　徐翰

2021 年 11 月 29 日

本书教学资源获取方式：

登录 heep.unipus.cn/product → 注册账号 → 再次登录 → 在搜索栏输入书名进行搜索 → 点击书名 → 点击"教学资源" → 下载教学资源。

Contents

Unit 1
China's development in aerospace

Learning objectives

Upon completion of this unit, you will be able to:

- clarify the history and future of Tiangong space station;
- illustrate the plan of China's astronautic development;
- evaluate China's role in the world's aerospace;
- write a summary.

◉ Lead-in

During the past decades, China has made painstaking efforts in space development and also witnessed tremendous advances in the area. For instance, when the three astronauts of China's Shenzhou-13 mission entered Tiangong space station, the station became the focus of the world again. Besides the diverse experiments carried out by the crews, China is looking to develop alternatives for keeping Tiangong supplied, which is a great prospect for the future. China's outline for space plans to 2025 also takes this into consideration. What's the future of China's aerospace? Let's work and see!

◉ Intensive reading

Tiangong space station

1   The three astronauts of China's Shenzhou-13 mission entered the country's Tianhe core module, the core of Tiangong space station, on October 16, 2021, to kick off a six-month expedition in the fledgling orbital lab. The astronauts entered the station at 9:58 a.m., Beijing Time, about eight hours after launching into orbit from Jiuquan Satellite Launch Center (JSLC) in the Gobi Desert. Their arrival capped a smooth autonomous docking by the Shenzhou-13 spacecraft, and Tiangong space station also caught the eyes of the world once more.

2   Tiangong is a space station that the China Manned Space Agency (CMSA) is building in low Earth orbit. On April 29, 2021, China launched Tianhe, the first of the orbiting space station's three modules, and the country aims to finish building the station by the end of 2022. CMSA hopes to keep Tiangong inhabited continuously by three astronauts for at least a decade. The space station will host many experiments from both China and other countries.

3   Tiangong, which means "Heavenly Palace", consists of Tianhe, the main habitat for astronauts, and two modules dedicated to hosting experiments, Mengtian and Wentian, both of which are due to launch in 2022. Shenzhou spacecraft, launching from Jiuquan in the Gobi Desert, have sent crews of three astronauts to the space station, while Tianzhou cargo spacecraft have launched from Wenchang on the Chinese island of Hainan to deliver supplies and fuel to the station.

Tiangong space station specifications

4   Tiangong is smaller than the International Space Station (ISS), with only three modules compared with the current 16 modules on the ISS. Tiangong is also lighter than the ISS, which weighs about 450 metric tons following the recent addition of Russia's Nauka lab module. The 16.6-meter-long Tianhe module launched with a docking hub that allows it to receive Shenzhou and Tianzhou spacecraft, as well as welcome the two later lab modules. A large robotic arm will help position the Mengtian and Wentian modules and assist astronauts during spacewalks.

5 Tianhe is much larger than the Tiangong-1 and -2 space labs China launched in the last decade and nearly three times heavier, at 22 metric tons. The new Tiangong, visiting spacecraft and cargo spacecraft will expand the usable space for the astronauts so much that they'll feel as though "they will be living in a villa," compared with how little space was available on previous Chinese space labs, a chief designer of the space station said. Tianhe features regenerative 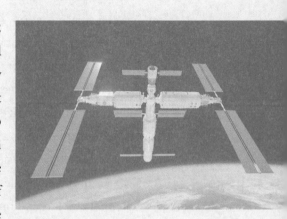 life support, including a way to recycle urine, to allow astronauts to stay in orbit for long periods. It is the main habitat for the astronauts and also houses the propulsion systems to keep the space station in orbit.

6 China has said it will take 11 launches to finish Tiangong: three module launches, four crewed missions and four Tianzhou spacecraft to supply cargo and fuel. The first three launches – Tianhe, Tianzhou-2 and Shenzhou-12 – have gone smoothly. Once completed, Tiangong will be joined by a huge, Hubble-like space telescope, which will share the space station's orbit and be able to dock for repairs, maintenance and upgrades. Named Xuntian, which translates as "survey the heavens," the space telescope will have a two-meter diameter mirror, like Hubble, but will have a field of view 300 times greater. Xuntian will aim to survey 40% of the sky over 10 years using its huge 2.5-billion-pixel camera. The space station could potentially be expanded to six modules, if everything goes according to plan. "We can further expand our current three-module space station combination into a four-module, cross-shaped combination in the future," the designer said. The Tianhe core module could then allow two more modules to join the orbital outpost.

Tiangong's project history

7 China embarked on a long journey to reach the point of building its space station. The project was first approved in 1992, after which the country set about developing the Shenzhou crew spacecraft and the Long March 2F rocket to send astronauts into space. Yang Liwei became China's first astronaut in space in October 2003 and made

China the third country in the world to independently send humans into orbit.

8   To be able to build and operate a crewed space station, China first needed to test out crucial space station systems, including life support and technologies for rendezvous and docking of spacecraft in orbit. To accomplish this, China launched the 8.5-metric-ton Tiangong-1 space lab in 2011, and subsequently sent the uncrewed Shenzhou-8 and the crewed Shenzhou-9 and -10 to join Tiangong-1 in orbit. The upgraded but similarly sized Tiangong-2 launched in 2016 and hosted the two-astronaut crew of Shenzhou-11 for just over a month, setting a new national record for human spaceflight mission duration.

9   As CMSA checked off these initial milestones, the agency was also focused on developing new, larger Long March heavy-lift rockets to make a space station possible. The Long March 5B rocket was designed specifically to launch the huge space station modules into low Earth orbit. In 2014, China completed its new coastal spaceport at Wenchang, specifically to launch these larger-diameter rockets, which need to be delivered by sea.

Tiangong's future

10 The first crewed missions – Shenzhou-12, -13 and -14 – will be for space station construction. A series of operational phase missions lasting six months each will begin in 2023. Crews will carry out experiments in areas such as astronomy, space medicine and life sciences, biotechnology, microgravity combustion and fluid physics, and space technologies. Tiangong will also temporarily host six astronauts during crew changeovers.

11 China is looking to develop alternatives for keeping Tiangong supplied. In January 2021, CMSA put out a call for proposals for low-cost, reliable cargo missions to Tiangong. The call was open to commercial companies, echoing NASA's Commercial Resupply Services contracts that provided opportunities to SpaceX.

12 It will be possible to spot Tiangong from Earth, just as it sometimes is with the ISS. Tiangong orbits at an altitude of between 340 to 450 kilometers above Earth and between 42 degrees north and south, and the space station is expected to be a fixture in the sky for at least a decade.

New words

module /ˈmɒdjuːl/ *n.* a unit of a spacecraft that can function independently of the main part（航天器上独立的）舱

expedition /ˌekspəˈdɪʃən/ *n.* an organized journey with a particular purpose, especially to find out about a place that is not well known 远征；探险

fledgling /ˈfledʒlɪŋ/ *adj.* recently formed 新形成的

cap /kæp/ *v.* to mean it is the final event in the series 使···结束

autonomous /ɔːˈtɒnəməs/ *adj.* having the ability to work and make decisions by yourself without any help from anyone else 自主的，有自主能力的

inhabit /ɪnˈhæbət/ *v.* to live in a particular place 居住在，栖居于

metric /ˈmetrɪk/ *adj.* using or relating to a system of measurement that uses meters, centimeters, liters, etc. 公制的，米制的

position /pəˈzɪʃən/ *v.* to put sb./sth. in a particular position 安置，放置

feature /ˈfiːtʃə/ *v.* to include a particular person or thing as a special feature 以···为特色

regenerative /rɪˈdʒenərətɪv/ *adj.* causing sth. to heal or become active again after it has been damaged or inactive 再生的；恢复的

urine /ˈjʊərən/ *n.* the waste liquid that collects in the bladder and that you pass from your body 尿，小便

propulsion /prəˈpʌlʃən/ *n.* the force that drives sth. forward 推动力

crew /kruː/ *v.* to be part of a crew（在飞机或船上）当工作人员

pixel /ˈpɪksəl/ *n.* any of the small individual areas on a computer screen, which together form the whole display 像素（组成屏幕图像的最小独立元素）

cross-shaped /ˌkrɒsˈʃeɪpt/ *adj.* shaped in the form of a cross 十字形的

outpost /ˈaʊtpəʊst/ *n.* a small military camp away from the main army, used for watching an enemy's movements, etc. 前哨（基地）

rendezvous /ˈrɒndəvuː/ *n.* an occasion when two spacecraft or military airplanes or vehicles meet, for example to move supplies from one to the other 会合，交会

spaceflight /ˈspeɪsflaɪt/ *n.* a flight beyond the Earth's atmosphere 宇宙飞行

duration /djʊˈreɪʃən/ *n.* the length of time that sth. lasts or continues 持续时间

microgravity /ˈmaɪkrəʊˌgrævəti/ *n.* very weak gravity, especially in a spacecraft orbiting the Earth 微重力

combustion /kəmˈbʌstʃən/ *n.* a chemical process in which substances combine with the oxygen in the air to produce heat and light 燃烧；燃烧过程

changeover /ˈtʃeɪndʒəʊvə/ *n.* a change from one system, or method of working to another 变更，改变，转变

fixture /ˈfɪkstʃə/ *n.* a thing that is fixed in a house or building 固定设施，固定装置

Phrases and expressions

kick off 开始（会议、活动等）	set about 开始，着手
catch the eyes of 吸引…的注意	test out 试验，检验
be dedicated to 致力于	check off 在…上打钩〔表示选择、完成
be due to 将要，预计	等〕
embark on 着手，开始	look to 展望
reach the point of 实现	put out 发布

Proper nouns

Tianhe core module 天和核心舱	Hubble space telescope 哈勃空间望
Tiangong space station 天宫空间站	远镜
Gobi Desert 戈壁沙漠	Xuntian space telescope 巡天空间望
Mengtian (lab module) 梦天（实验舱）	远镜
Wentian (lab module) 问天（实验舱）	Long March 2F rocket 长征二号F运
Tianzhou cargo spacecraft 天舟货运飞船	载火箭
Russia's Nauka lab module 俄罗斯科学	Long March 5B rocket 长征五号B运
号实验舱	载火箭

Technical terms

low Earth orbit 近地轨道	microgravity combustion and fluid
docking hub 对接枢纽	physics 微重力燃烧科学和流体物
space lab 空间实验室	理学

Background information

Shenzhou-13 神舟十三号

Shenzhou was approved in 1992 as part of the Chinese space program Project 921. The China Academy of Space Technology developed new technology on Shenzhou-13 in 2021 for the purpose of the radial docking to the bottom of Tianhe, requiring maintenance of continuous altitude and orbit control. This type of docking differed from the front and rear dockings used by the previous Shenzhou-12 as well as Tianzhou-2 and -3 missions, creating a logistic chain of space infrastructure that includes the core module, manned and cargo spacecraft.

Jiuquan Satellite Launch Center (JSLC) 酒泉卫星发射中心

Jiuquan Satellite Launch Center is a Chinese space vehicle launch facility (spaceport) located in the Gobi Desert. It is also called the Dongfeng Aerospace City. Although part of the facility is geographically located within Ejina Banner of Inner Mongolia's Alxa League, it is named after the nearest city, Jiuquan in Gansu Province. The launch center straddles both sides of the Ruoshui River. It was founded in 1958, the first of China's four spaceports. JSLC is usually used to launch vehicles into lower and medium orbits with large orbital inclination angles, as well as testing medium- to long-range missiles.

China Manned Space Agency (CMSA) 中国载人航天

China Manned Space Agency, or China Manned Space (CMS), is a Chinese agency responsible for the administration of the country's manned space program. Its functions include development strategy, planning, overall technology, research and production, infrastructure construction, flight mission organization and implementation, utilization and promotion, international cooperation and news-releases, etc.

International Space Station (ISS) 国际空间站

The International Space Station (ISS) is a modular space station (habitable artificial satellite) in low Earth orbit. It is a multinational collaborative project involving five participating space agencies: NASA (America), Roscosmos (Russia), JAXA (Japan), ESA (Europe), and CSA (Canada). The ownership and use of the space station is established by intergovernmental treaties and agreements. The station serves as a microgravity and space environment research laboratory in which scientific research is conducted in astrobiology, astronomy, meteorology, physics, and other fields. The ISS is suitable for testing the spacecraft systems and equipment required for possible future long-duration missions to the Moon and Mars.

NASA (National Aeronautics and Space Administration) 美国国家航空航天局

The National Aeronautics and Space Administration is America's space agency which studies the Earth, including its climate, the Sun, and the solar system and beyond. NASA also leads a Moon to Mars exploration approach, which includes working with US industry, international partners, and academia to develop new technology, and send science research and soon humans to explore the Moon on Artemis missions that will help prepare for human exploration of the Red Planet.

SpaceX 美国太空探索技术公司

SpaceX, short for Space Exploration Technologies Corp., is an American aerospace manufacturer, space transportation services and communications corporation headquartered in California. It was founded in 2002 by Elon Musk with the goal of reducing space transportation costs to enable the colonization of Mars. SpaceX manufactures the Falcon 9 and Falcon Heavy launch vehicles, several rocket engines, crew spacecraft and Starlink communications satellites.

Difficult sentences

1 **The three astronauts of China's Shenzhou-13 mission entered the country's Tianhe core module, the core of Tiangong space station, on October 16, 2021, to kick off a six-month expedition in the fledgling orbital lab. (Para. 1)**

Shenzhou-13 transported three crew members to Tianhe who went onboard the main section of Tiangong on October 16, 2021, which was the start of their six-month adventure in the brand-new space station.

2 **Their arrival capped a smooth autonomous docking by the Shenzhou-13 spacecraft, and Tiangong space station also caught the eyes of the world once more. (Para. 1)**

Shenzhou-13 made the successful completion of a smooth automated connection to Tiangong, bringing the space station to the attention of the world again.

3 **Tiangong, which means "Heavenly Palace", consists of Tianhe, the main habitat for astronauts, and two modules dedicated to hosting experiments, Mengtian and Wentian, both of which are due to launch in 2022. (Para. 3)**

Tiangong, "Heavenly Palace" in Chinese, is comprised of Tianhe, the main living place for astronauts, and two lab modules, namely Mengtian and Wentian, which will be used for doing experiments. These two modules are expected to lift off in 2022.

4 **Shenzhou spacecraft, launching from Jiuquan in the Gobi Desert, have sent crews of three astronauts to the space station, while Tianzhou cargo spacecraft have launched from Wenchang on the Chinese island of Hainan to deliver supplies and fuel to the station. (Para. 3)**

Lifting off from Jiuquan in the Gobi Desert, Shenzhou spacecraft brought groups of three astronauts to the space station, and Tianzhou cargo spacecraft lifted off from Wenchang, Hainan, transporting supplies and fuel to the station.

5 **The 16.6-meter-long Tianhe module launched with a docking hub that allows it to receive Shenzhou and Tianzhou spacecraft, as well as welcome the two later lab modules. (Para. 4)**

Tianhe module is 16.6 meters long and it lifted off with facilities, allowing it to connect safely to Shenzhou and Tianzhou spacecraft, and also to connect to the two other modules for experiments which will be subsequently added.

6   **The new Tiangong, visiting spacecraft and cargo spacecraft will expand the usable space for the astronauts so much that they'll feel as though "they will be living in a villa," compared with how little space was available on previous Chinese space labs, a chief designer of the space station said. (Para. 5)**

The new Tiangong, visiting spacecraft and cargo spacecraft will enlarge the space for astronauts. The new station is so spacious that astronauts feel comfortable as if they were living in a villa, which is unlike the quite small space labs, according to a chief designer of the project.

7   **Once completed, Tiangong will be joined by a huge, Hubble-like space telescope, which will share the space station's orbit and be able to dock for repairs, maintenance and upgrades. (Para. 6)**

When the assembly of Tiangong is finished, a huge space telescope similar to Hubble will be located nearby. The telescope will be on the same orbit as the space station and be able to dock at the station to be repaired, maintained and upgraded.

8   **The project was first approved in 1992, after which the country set about developing the Shenzhou crew spacecraft and the Long March 2F rocket to send astronauts into space. (Para. 7)**

In 1992, the government first reached a decision on space exploration. Since then, the country started to design and build Shenzhou spacecraft for astronauts and the Long March 2F rocket to transport them off Earth.

9   **To be able to build and operate a crewed space station, China first needed to test out crucial space station systems, including life support and technologies for rendezvous and docking of spacecraft in orbit. (Para. 8)**

After testing out the important space station systems, China would be able to build and operate a manned space station. The space station systems include life support and technologies for meeting and connecting of spacecraft in orbit.

10   **The upgraded but similarly sized Tiangong-2 launched in 2016 and hosted the two-astronaut crew of Shenzhou-11 for just over a month, setting a new national record for human spaceflight mission duration. (Para. 8)**

Though Tiangong-2 had improved technology, the size remained almost the same. It lifted off in 2016 and Shenzhou-11 docked with it with two astronauts aboard for just over a month. This was a new national record for Chinese astronauts staying in space for so long.

11 **The call was open to commercial companies, echoing NASA's Commercial Resupply Services contracts that provided opportunities to SpaceX. (Para. 11)**

The request was available to any commercial company, similar to the contracts of Commercial Resupply Services of NASA. These contracts provided opportunities to SpaceX, a famous space technology company in America.

12 **Tiangong orbits at an altitude of between 340 to 450 kilometers above Earth and between 42 degrees north and south, and the space station is expected to be a fixture in the sky for at least a decade. (Para. 12)**

Tiangong travels at the height above Earth between 340 to 450 kilometers and between 42 degrees north and south. It should be a permanent station in the space for at least ten years.

Exercises

I Fill in the blanks with the words and expressions given below, and change the forms when necessary.

autonomous	regenerative	rendezvous	kick off	propulsion
feature	fixture	combustion	position	test out

1 Many of the hotels _____ in the brochure offer special deals for weekend breaks.

2 The high temperature necessary for the reaction is obtained by the _____ of coke.

3 His work is to _____ new designs of cars before they are put on the market.

4 The Chinese e-commerce giant reported its accumulated sales rose to 349.1 billion yuan during the Singles' Day shopping carnival, which _____ at 8 p.m. on Oct. 31.

5 During this promotional gala, the company's logistics applied nearly 400 _____ delivery vehicles in more than 25 cities in China.

6 As a matter of fact, we have come to a point where agroforestry, _____ agriculture, and technology must be combined in order to save us from the destruction.

7 The spacecraft entered a low Earth orbit and began to execute the rapid
   _____ and docking procedures with Tianhe, the space station's core
   module.

8 A Tianzhou cargo spacecraft has two parts – a cargo cabin and a(n)
   _____ section – and can transport up to 6.9 metric tons of supplies.

9 Large television screens were _____ at either end of the stadium to video
   the whole event.

10 He has stayed with us so long that he seems to have become a permanent
    _____.

|| Translation

Part A   Translate the following paragraph into Chinese.

As CMSA checked off these initial milestones, the agency was also focused on
developing new, larger Long March heavy-lift rockets to make a space station
possible. The Long March 5B rocket was designed specifically to launch the huge
space station modules into low Earth orbit. In 2014, China completed its new coastal
spaceport at Wenchang, specifically to launch these larger-diameter rockets, which
need to be delivered by sea.

Part B   Translate the following paragraph into English.

2021年9月，三名中国宇航员成功完成了三个月的太空任务，期间他们在天和核心舱
工作，并进行了两次太空行走以安装设备。一个月后，即10月16日，中国发射了神
舟十三号载人飞船，与正在近地轨道运行的中国天宫空间站天和核心舱交会对接。

||| Write a summary of the text in 120 words.

Directions: A summary should be written in your own words. It should contain only
the ideas of the original text. Do not insert any of your own opinions, interpretations,
deductions or comments into the summary.

IV Read the passage, select one word for each blank from a list of choices in the bank, and change the form when necessary. You may not use any of the words in the bank more than once.

> assemble   crew   psychologically   module   launch
>
> approximately   carry out   blast off   conduct   deploy

China on Saturday 1) _____ a rocket carrying three astronauts – two men and one woman – to the core 2) _____ of a future space station where they will live and work for six months, the longest orbit for Chinese astronauts. A Long March 2F rocket carrying the Shenzhou-13 spacecraft 3) _____ from Jiuquan Satellite Launch Center in the northwestern province of Gansu at 00:23 a.m., Oct. 16, 2021.

China began constructing the space station in this April with the launch of Tianhe, the first and largest of the station's three modules. Shenzhou-13 is the second of four 4) _____ missions needed to complete the space station 5) _____ by the end of 2022. In the latest mission, astronauts will 6) _____ tests of the key technologies and robotics on Tianhe needed to 7) _____ the space station, verify onboard life support systems and 8) _____ a host of scientific experiments. Besides, the long-term stay in orbit for six months will exact higher demands, both physically and 9) _____. After this, China plans to 10) _____ six more missions, including deliveries of the second and third space station modules and two final crewed missions. With the ISS set to retire in a few years, China's space station will become the only one in Earth's orbit.

It turned out to be a great success since China became the third country to put a man in space with its own rocket, in October 2003, following the Soviet Union and the United States.

V Critical thinking

> **Work in groups and discuss the following questions.**
> Tiangong space station is the symbol of independent research and development in China's aerospace. Do you think Tiangong is indispensable in the development of China's aerospace, or even the world's aerospace, with the ISS set to retire in a few years? Why or why not?

◉ **Further reading**

China outlines space plans to 2025

1   China's space administration has outlined its priorities in space science, technology, application and exploration for the coming years. Lunar, interplanetary and near-Earth asteroid missions, space station construction, a national satellite internet project and developing heavy-lift launch vehicles and reusable space transportation systems are noted as major projects for the period 2021-2025.

2   One official from China National Space Administration (CNSA) laid out the main activities and focus of the country's civilian space endeavors in a press conference on June 12, 2021. Boosting innovation, supporting economic and social development and engaging in international cooperation were noted as major objectives.

3   In lunar exploration the Chang'e-6 sample return and complex Chang'e-7 south pole mission are to be conducted during China's "14th Five-Year Plan" period. Chang'e-8, to include in-situ resource utilization and 3D-printing technology tests, will follow. All missions will form part of the first phase of the International Lunar Research Station (ILRS) project with Russia.

4   CNSA is also looking to build on the recent success of the country's first independent interplanetary expedition with the Tianwen-1 Mars orbiter and Zhurong rover. Development of a Mars sample return mission and a Jupiter probe for launches around 2028 and 2030 respectively are noted as follow-up projects. "So far, our knowledge of the Jupiter system is very superficial, and the detections performed are also very limited," said the chief designer of Tianwen-1. "The Jupiter system offers a large number of opportunities for scientific discoveries." One proposal for the mission includes a landing on the Galilean moon Callisto. The chief designer also stated that technology breakthroughs are needed for missions. "Everyone knows that, so far, no country in the world has been able to carry out a sample and return from Mars, because it is too technically difficult." China performed a complex lunar

sample return in late 2020 but he noted that the challenges of launching samples from the Mars surface were different to that of the moon.

5　Launching around 2025 will be a near-Earth asteroid sample return mission to small body 469219 Kamo'oalewa. The mission was previously targeting a 2024 launch, with the secondary target following the delivery of samples to Earth last understood to be main-belt comet 311P/PANSTARRS. Not mentioned are a pair of probes to launch for the head and tail of the heliosphere, which is however led by figures from the Chinese Academy of Sciences (CAS). In human spaceflight China aims to complete the construction of its three-module space station by the end of 2022. The Tianhe core module launched late April and is currently hosting its first crew.

6　CNSA also aims to enhance satellite application capabilities over the next five years. Goals include improving national civil space infrastructure and supporting ground facilities and enhancing Earth observation, communications and broadcasting, and navigation and positioning capabilities, as well as promoting and supporting downstream applications to boost economic development. China recently established a company to oversee development of a 13,000-satellite constellation for satellite internet.

7　Expanding international exchanges and cooperation is another major strand. Citing guiding principles of equality, mutual benefit, the peaceful use of outer space and inclusive development, one official from CNSA noted projects include the ILRS, a second Sino-Italian seismo-electromagnetic satellite, a follow-up to the China-Brazil Earth Resources Satellite (CBERS) program and the China-France SVOM X-ray space telescope.

8　The upcoming Chang'e and asteroid missions will also include international payloads. The future lunar, Mars and Jupiter missions will be opened to international cooperation, according to the official.

9　Other cooperation activities include promoting the construction of the "Belt and Road" spatial information corridor, the BRICS remote sensing satellite constellation and jointly responding to the common challenges of global climate change and dangers of near-Earth asteroids. The under-construction Wenchang International

Aerospace City is expected to be developed as a hub for international scientific research, academic exchanges, exhibitions and training.

10 Details on launch vehicle technology were not offered, other than underlining its fundamental importance to progress in aforementioned missions. The development of separate super heavy-lift launchers in March was approved for infrastructure and crewed flights. China's main space contractor CASC is developing a first vertical takeoff, vertical landing launcher in the Long March 8 and working on a "reusable experimental spacecraft" widely held to be a spaceplane.

11 Satellite launches are a priority, too. China attempts to catch up with the West's satellite infrastructure. China launched a rocket from a mobile platform in the Yellow Sea in June 2019, sending five commercial satellites and two others containing experimental technology into orbit. The feat meant China is only the third country after the US and Russia to master sea launches.

12 The speed at which China is surpassing each technological hurdle spotlights how the space was viewed as vital for boosting the economy and promoting high-end industry. "They see space as a very important driver for growth and competitiveness going forward," said a journalist specializing in China's space program.

13 China already has the largest filled-aperture radio telescope in the world, which measures just over 500 meters across. More than visiting Mars, China plans to send probes to asteroids and even Uranus. It also aims to build a scientific research station in the moon's southern polar region, and establish its own sophisticated large-scale space station within 10 years. "They have a long-term set of goals and work deliberately and systematically to achieve those goals," said Kathy Laurini, NASA's senior advisor.

14 There are some challenges that the China's space sector is facing, but there are more opportunities for new developments. The solutions to the various issues, which in time will be developed, will provide stepping stones for future applications and enterprises that will benefit societies across the world.

New words

interplanetary /ˌɪntəˈplænɪtəri◂/ *adj.* situated or relating to travel between planets 行星间的, 行星际的

asteroid /ˈæstərɔɪd/ *n.* a small planet orbiting the Sun 小行星

Mars /mɑːz/ *n.* the planet in the solar system that is fourth in order of distance from the Sun 火星

orbiter /ˈɔːbɪtə/ *n.* a spacecraft designed to go into orbit, especially one that does not subsequently land 轨道飞行器

Jupiter /ˈdʒuːpɪtə/ *n.* the planet in the solar system that is fifth in order of distance from the Sun 木星

comet /ˈkɒmɪt/ *n.* an object in space like a bright ball with a long tail, that moves around the Sun 彗星

heliosphere /ˈhiːliəʊsfɪə/ *n.* the region surrounding the Sun where the solar wind has a significant influence 日球层, 太阳风层

downstream /ˌdaʊnˈstriːm◂/ *adj.* situated or moving in the direction in which a stream or river flows 在下游的, 顺流而下的

constellation /ˌkɒnstəˈleɪʃn/ *n.* a group of stars forming a recognizable pattern that is traditionally named after its apparent form or identified with a mythological figure 星座

strand /strænd/ *n.* one of the parts of a story, idea, plan, etc. (故事、观点、计划等的) 部分

mutual /ˈmjuːtʃuəl/ *adj.* (of a feeling or action) experienced or done by each of two or more parties toward the other or others; (of two or more people) having the same specified relationship to each other 相互的, 彼此的; 共同的, 共有的

payload /ˈpeɪləʊd/ *n.* the amount of goods or people that a vehicle, such as a truck or aircraft, can carry 有效载荷

aforementioned /əˈfɔːmenʃənd/ *adj.* denoting a thing or person previously mentioned 前面提到的, 上述的

contractor /kənˈtræktə/ *n.* a person or firm that undertakes a contract to provide materials or labor to perform a service or do a job 承包商, 承包公司

feat /fiːt/ *n.* an achievement that requires great courage, skill, or strength 壮举, 功绩

surpass /səˈpɑːs/ *v.* to be even better or greater than someone or something else 超过; 胜过

Uranus /ˈjʊərənəs/ *n.* the planet in the solar system that is seventh in order of distance from the Sun 天王星

deliberately /dɪˈlɪbərətli/ *adv.* in a careful and unhurried way 慎重地, 从容不迫地

Phrases and expressions

be noted as 被列为; 以…闻名	high-end industry 高端工业
catch up with 追上, 赶上	stepping stone 垫脚石, 跳板

Proper nouns

China National Space Administration
  (CNSA) 中国国家航天局
14th Five-Year Plan 十四五规划
International Lunar Research Station
  (ILRS) 国际月球科研站
Galilean moon Callisto 伽利略卫星卡利
  斯托
Chinese Academy of Sciences (CAS) 中
  国科学院
China–Brazil Earth Resources Satellite
  (CBERS) 中巴地球资源卫星

SVOM (Space Variable Objects
  Monitor) 天基多波段空间变源监
  视器
Belt and Road 一带一路
BRICS 金砖国家（因Brazil、Russia、
  India、China、South Africa五个国
  家的英文单词首字母缩写与英语单
  词brick<砖>相似而得名）
Wenchang International Aerospace
  City 文昌国际航天城
CASC (China Aerospace Science and
  Technology Corporation) 中国航天
  科技集团有限公司

Technical terms

in-situ resource utilization 原位资源利用	电磁（监测试验）卫星
main-belt comet 主带彗星	filled-aperture radio telescope 连续孔
seismo-electromagnetic satellite 地震	径射电望远镜

Exercises

For each of the following unfinished statements or questions, choose the most appropriate answer from A, B, C, or D according to the text.

1 Which of the following items is NOT included in the major projects for the period 2021-2025?

A. Lunar, interplanetary and near-Earth asteroid missions.

B. A national satellite internet project.

C. Developing light-lift launch vehicles.

D. The construction of a space station.

2 According to the chief designer of Tianwen-1, why hasn't any country carried out a sample and return from Mars so far?

A. Because the detections that we performed are very limited.

B. Because our knowledge of the Jupiter system is very superficial.

C. Because of technical limitations.

D. Because of the lack of funds.

3 What is so special about the Long March 8?

A. It will be equipped with separate super heavy-lift launchers.

B. It will be able to take off and land vertically.

C. It played an important role in aforementioned missions.

D. Its mission will include international payloads.

4 _____ marked China's mastery on sea launches.

A. The launch of Long March 8

B. The construction of Wenchang International Aerospace City

C. The development of a "reusable experimental spacecraft"

D. The launch of a rocket from a mobile platform in the Yellow Sea

5 According to Kathy Laurini, China will achieve its goals by _____.

A. sending the probes to asteroids

B. building a scientific research station

C. establishing its own sophisticated large-scale space station

D. working gradually and systematically

Questions for discussion

1 What missions will form part of the first phase of the ILRS project?

2 What are the goals of CNSA that aims to enhance satellite application capabilities?

3 What guiding principles does China follow in its international exchanges and cooperation with other countries?

4 What function is the Wenchang International Aerospace City expected to have?

5 What are China's plans for the next 10 years?

⊙ Practical writing

How to write a summary

To summarize is to condense a text to its main points in your own words. You do not need to include every detail. Instead, you should extract only those elements that you consider most important – the main idea (or thesis) and its essential supporting points.

Many students make the mistake of confusing a summary with an analysis. They are not the same thing. An analysis is a discussion of ideas. A summary, on the other hand, does not require you to respond to the ideas in a text. When analyzing a piece of writing, you generally summarize the content briefly in order to establish for the reader the ideas that your essay will then go on to analyze.

Be sure to remember that a summary is not an outline of the points. Instead, it is focusing on the main ideas or arguments of the text. It is a reconstruction of the major point or points of development of a text, beginning with the thesis or main idea, followed by the points or details that support or elaborate on that thesis or idea.

For the sake of clarity, a summary should present the author's points in a straightforward structure. In order to write a good summary, you may have to gather minor points or components of an argument from different places in the text in order to summarize the text in an organized way. A point made in the beginning of an essay and then one made toward the end may need to be grouped together in your summary to concisely convey the argument that the author is making.

Depending on the length and complexity of the original text as well as your purpose in using the summary, a summary can be relatively brief (a short paragraph or even a single sentence), or quite lengthy (several paragraphs or even an entire paper).

A good summary should be **comprehensive, concise, coherent,** and **independent**. These qualities are explained as follows.

A summary must be comprehensive: You should isolate all the important points in the original passage and note them down in a list. Review all the ideas on your list, and include in your summary all the ones that are necessary to the author's development of the thesis or main idea.

A summary must be concise: Eliminate repetitions in your list, even if the author restates the same points. Your summary should be considerably shorter than the original. It is expected that you create an overview; therefore, you don't need to include every repetition of a point or every supporting detail.

A summary must be coherent: It should make sense as a piece of writing in its own right; it should not merely be taken directly from your list of notes.

A summary must be independent: You are not being asked to imitate the author of the text you are writing about. On the contrary, you are expected to maintain your own voice throughout the summary. Don't simply quote the author; instead use your own words to express your understanding of what you have read. After all, your summary is based on your interpretation of the author's points or ideas. However, you should be careful not to create any misrepresentation or distortion by introducing comments or criticisms of your own.

Once you are certain that your summary is accurate, you should revise it for style, diction, grammar, and punctuation.

Exercise

Write a one-paragraph summary of the text "China outlines space plans to 2025". Remember that the summary should be comprehensive, concise, coherent and independent.

Unit **2**
World space exploration

Learning objectives

Upon completion of this unit, you will be able to:

- understand how space travel develops worldwide;
- get some insights into human cooperation and exploration of space;
- evaluate how we can apply space science in our daily life in an optimal way;
- use appropriate language in academic writing.

◉ Lead-in

Have you ever wondered how far humans have reached in deep space? Has it ever occurred to you that you can travel to space someday? Although space travel might be currently out of reach for common people, modern science and technology have made it possible in the not too distant future. Achievements in the development of lunar expeditions not only push forward the development of space science, but also enhance our confidence to learn more about the universe.

⊙ Intensive reading

Space: How far have we gone?

Who has traveled to space?

1　Space flight is now an admirable industry. Humanity's first space explorer, Soviet cosmonaut Yuri Gagarin, orbited around the globe on April 12, 1961, more than half a century ago, when Britain remained a colonial power and people were still using halfpennies to buy their fish and chips.

2　Since then, more than 550 people have blasted themselves into deep black space, although not all agree on how far up you need to go until you hit space, so there is no internationally accepted figure. Only one tenth of those have been women, in big part due to sexist policies by NASA and Russia's Roscosmos space agency.

Where have we been in space?

3　The Soviet Union pulled ahead with the first space walks, but then US president John F. Kennedy's announcement that America would put a man on the moon by the end of the 1960s focused the space race exactly on that goal. Apollo 11 touched down on our dusty grey neighbor on July 20, 1969.

4　A total of 12 men walked on the moon over the next few years, all Americans, but no one has been back there since 1972. In fact, no one has left the outskirts of the Earth since then.

5　We imagine astronauts floating in free space or bouncing on the moon, yet the majority of those lucky enough have instead spun around in low Earth orbit – between 99 and a few hundred kilometers high. That's where the vast array of communications and navigation satellites live, speeding at thousands of kilometers an hour to avoid plummeting back to Earth.

What happens to the body in space?

6　Until we properly understand how weightlessness

affects humans, we won't be able to send this era's pioneers further afield to places such as Mars or asteroids. Scott Kelly, a US former fighter pilot and long-time NASA astronaut, spent a year bouncing around the capsules of the International Space Station (ISS) in an attempt to understand the long-term impact of space flight. He doesn't hold the record for the most extended journey into the void – that is claimed by Gennady Padalka, who spent two and half years of his life up there on several missions – but the Kelly experiment had a natural advantage over others: He has a twin.

7 Comparing their bodies throughout, scientists were able to assess how bones, muscles and other parts of the body deteriorate in space. There is even a gym on the ISS where astronauts can keep their muscles from slowly wasting away. But they need to wear a harness to keep them from floating off the treadmill. One big issue is that eye problems develop, but Kelly found his body recovered fast on return. He and his twin seemed in similar shape – good news for future deep space missions.

How much does it cost to send them up?

8 Astronomical. The ISS is the most expensive machine ever constructed with a price tag at around $150bn. NASA's space shuttle program, which kicked off in the early 1970s by promising safe and affordable access to space, hoped to cost just a few tens of million dollars per launch. But as the shuttle was thrown away in 2011, the agency estimated the total cost at $209bn – nearly $1.6bn per flight.

9 Following the big fight over the shuttle, which looked fantastic but also restricted space adventuring to Earth's orbit as well as costing a fortune, the US took a back seat in launches. Most astronauts are then sent by the Russian space agency, which sells round-trip rides on its Soyuz spacecraft for between $21m and $82m.

Is human space flight worth the cost?

10 Anyone involved in space travel will scoff at this, but it's a good question, and space agencies often don't communicate their achievements enough. Almost every sector of human progress has benefited from sending people into space. Just the act of attempting the feat forced scientists to invent new systems. The Apollo Guidance Computer was a predecessor to the microcomputer now found in all smartphones. Clothes are more fire-resistant because of research on space fires. Remotely

monitoring the health of astronauts has led to revolutionary systems for helping patients on Earth. Diseases behave and develop differently in microgravity, which assists scientists in finding cures.

11　Others say paying for human space flight pumps money into the economy, arguing that spin-off companies from space research and a growing commercial space industry generate 7 to 14 times the cost of missions.

12　And NASA is not spending nearly as much as it used to. About $19bn is spent by the US government on its budget, roughly half a percent of all federal spending. During the early Apollo program, that was between 4% and 5%.

How strong is space cooperation between countries?

13　The first space race was part of the chest-beating of the Cold War, but since then human space exploration has been more about countries working together than against each other. The ISS is a massive collaboration between five space agencies (NASA, Roscosmos, Japan's JAXA, the pan-European agency ESA and Canada's CSA) and was assembled over a period of 13 years from 1998, slowly adding capsules like Lego.

14　The future of any effective human space flight is certainly likely to be cooperative rather than competitive. Since 2011, national space agencies in 14 countries have attempted to coordinate their dreams into a single vision. The most recent plan, published in January 2018, said they had agreed to "expand human presence into the solar system, with the surface of Mars as a common driving goal".

We're off to Mars? Hurray!

15　Don't start the countdown just yet. To get to Mars, most people in the human space flight community feel we need to first go back to the moon. "It's the only logical step," says Ian Crawford, professor of planetary science and astrobiology at Birkbeck, University of London. "I'm all in favor of sending people to Mars, but the technology, the competence, the experience – I think they're still out of reach."

16　The moon has several advantages. It's only three days away, rather than a several-month round trip to Mars, and has been considered as an ideal location for a research station similar to the one in Antarctica. From their laboratory, scientists could study the impact of radiation exposure and near-weightlessness on the body at a closer distance to Earth, but still within deep space, all while preparing for trips further afield.

New words

cosmonaut /ˈkɒzmənɔːt/ *n.* an astronaut from the former Soviet Union（苏联的）宇航员

orbit /ˈɔːbɪt/ *v.* to move around sth. in a continuous, curving path 环绕…的轨道运行

blast /blɑːst/ *v.* to send or direct sth. out with great force 发射；喷射

bounce /baʊns/ *v.* to jump up and down on sth. 上下跳动，蹦跳

spin /spɪn/ *v.* to turn round quickly 快速旋转

array /əˈreɪ/ *n.* a group or collection of things or people, often one that is large or impressive 大堆；大批；大量

afield /əˈfiːld/ *adv.* to or in places that are not near 去远处；在远方

capsule /ˈkæpsjuːl/ *n.* the part of a spacecraft in which people travel and that often separates from the main rocket 太空舱

void /vɔɪd/ *n.* a large empty area of space 空间

deteriorate /dɪˈtɪəriəreɪt/ *v.* to become worse 变坏；恶化

harness /ˈhɑːnɪs/ *n.* a set of strips of leather, etc. for fastening sth. to a person's body or to keep them from moving off or falling（起固定或保护作用的）系带；保护带

astronomical /ˌæstrəˈnɒmɪkəl◂/ *adj.* (of an amount, a price, etc.) extremely large or high（数量、价格等）极其巨大的，天文数字的

predecessor /ˈpriːdɪsesə/ *n.* a thing, such as a machine, that has been followed or replaced by sth. else; a person who did a job before sb. else 前身；前任

astrobiology /ˌæstrəʊbaɪˈɒlədʒi/ *n.* the branch of biology that investigates the possibility of life elsewhere in the universe 天体生物学

Phrases and expressions

fish and chips 炸鱼加薯条
blast sb. into space 把某人送入太空
pull ahead 领先
touch down 着陆，降落
waste away 日渐衰弱，日益消瘦

take a back seat 处于次要地位；退居二线
round-trip ride 往返行程
pump money into 注入大量资金
spin-off company 衍生公司

Proper nouns

Yuri Gagarin 尤里·加加林（苏联宇航员）.

John F. Kennedy 约翰·F. 肯尼迪（美国第35任总统）

Apollo 11 阿波罗11号（美国国家航空航天局的阿波罗计划中的第五次载人航天任务）

Scott Kelly 斯科特·凯利（美国宇航员）

Gennady Padalka 根纳季·帕达尔卡（俄罗斯宇航员）

Soyuz spacecraft 联盟号宇宙飞船

Apollo Guidance Computer 阿波罗导航计算机

Technical terms

space flight 太空飞行

navigation satellite 导航卫星

weightlessness 失重状态

space shuttle 航天飞机

microcomputer 微型计算机

solar system 太阳系

planetary science 行星科学

radiation exposure 辐射照射

near-weightlessness 近似失重的状态

Background information

Roscosmos 俄罗斯联邦航天局

Roscosmos is a state corporation of the Russian Federation responsible for space flights, cosmonautics programs, and aerospace research.

JAXA 日本宇宙航空研究开发机构

The Japan Aerospace Exploration Agency (JAXA) is the Japanese national aerospace and space agency, formed in October 2003 through the merger of three previously independent organizations. It is responsible for research, technology development, launch of satellites into orbit, and many more advanced missions.

ESA 欧洲航天局

The European Space Agency (ESA), established in 1975, is an intergovernmental organization dedicated to space exploration. It consists of twenty-two member states, headquartered in Paris.

CSA 加拿大航天局

The Canadian Space Agency (CSA), the space agency of Canada, was established in 1989 by the Canadian Space Agency Act which was proclaimed in 1990. The agency is responsible for managing all of Canada's civil space-related activities.

Difficult sentences

1 **Humanity's first space explorer, Soviet cosmonaut Yuri Gagarin, orbited around the globe on April 12, 1961, more than half a century ago, when Britain remained a colonial power and people were still using halfpennies to buy their fish and chips. (Para. 1)**

   Yuri Gagarin, a Soviet cosmonaut, traveled around Earth in space for the first time in history on April 12, 1961, over fifty years ago. Here it suggests that the Soviet Union explored space a long time ago. At that time Britain was still a colonial power (with many countries under its rule), and halfpennies (old coins) were spent on daily food.

   A halfpenny was a British coin in use until 1969, worth half an old penny. There were 480 halfpennies in an old English pound.

2 **Since then, more than 550 people have blasted themselves into deep black space, although not all agree on how far up you need to go until you hit space, so there is no internationally accepted figure. (Para. 2)**

   From that time until now, over 550 people have been sent into space. But it is hard to get the exact number as to how many people have gone into space because in the world there is no agreement on the distance you should travel until you leave Earth and reach space.

3 **Apollo 11 touched down on our dusty grey neighbor on July 20, 1969. (Para. 3)**

   Apollo 11 landed on the moon on July 20, 1969. The phrase "our dusty grey neighbor" here refers to "the moon" indirectly.

4 **That's where the vast array of communications and navigation satellites live, speeding at thousands of kilometers an hour to avoid plummeting back to Earth. (Para. 5)**

   A large number of communications and navigation satellites stay in low Earth orbit, and move quickly at thousands of kilometers an hour so as not to suddenly fall back to Earth.

5 **Others say paying for human space flight pumps money into the economy, arguing that spin-off companies from space research and a growing commercial space industry generate 7 to 14 times the cost of missions. (Para. 11)**

Others think that paying for human space flight puts a lot of money into the economy. The evidence is that an increasing number of companies related to space research and commercial space industry produce 7 to 14 times the amount of money spent on missions.

6 **The first space race was part of the chest-beating of the Cold War, but since then human space exploration has been more about countries working together than against each other. (Para. 13)**

The first space race was part of the fierce competition in the Cold War, but from then on, countries have been willing to cooperate in space exploration rather than compete against each other.

The word "chest-beating" in the sentence means trying to see who is the most dominant, or the most powerful. It originally comes from gorillas that beat their chests to frighten off potential threats.

7 **The ISS is a massive collaboration between five space agencies (NASA, Roscosmos, Japan's JAXA, the pan-European agency ESA and Canada's CSA) and was assembled over a period of 13 years from 1998, slowly adding capsules like Lego. (Para. 13)**

The ISS is a huge cooperation between five space organizations. It was built in stages during a span of more than 13 years which began in 1998, and was pieced together adding sections step by step.

8 **It's only three days away, rather than a several-month round trip to Mars, and has been considered as an ideal location for a research station similar to the one in Antarctica. (Para. 16)**

It takes only three days to send people to the moon, rather than several months needed to go to Mars and then return. The moon is believed to be an ideal place for a research station, just like the research base in Antarctica.

Exercises

I **Fill in the blanks with the words and expressions given below, and change the forms when necessary.**

> blast into   deteriorate   fire-resistant   microgravity   navigation
>
> waste away   astronomical   capsule   radiation   collaboration

1 The government worked in close _____ with teachers on the curriculum.
2 The President's science adviser was asked about the benefits of doing experiments in weightlessness and _____.
3 She says the boat would be shaped like a(n) _____, powered by a small nuclear engine.
4 We don't have that problem anymore, because we have _____ systems like GPS.
5 The area should be dry, cool, and especially _____ to prevent spoilage and the swelling of canned goods.
6 Japan has been trying to bring its nuclear crisis under control, but the spread of _____ has raised concerns about the safety of the nuclear energy.
7 The Federal Reserve reassured investors it was ready to adjust its policy of winding down its crisis-era stimulus program if the economy _____.
8 You get thinner every time I see you, Sara – you're _____!
9 Financial transactions, sales figures, inventory tallies – the number of data items can be _____.
10 Apollo 11 was _____ space from the Kennedy Space Center, Florida, on July 16, 1969.

II Translation

Part A   Translate the following paragraph into Chinese.

Almost every sector of human progress has benefited from sending people into space. Just the act of attempting the feat forced scientists to invent new systems. The Apollo Guidance Computer was a predecessor to the microcomputer now found in all smartphones. Clothes are more fire-resistant because of research on space fires. Remotely monitoring the health of astronauts has led to revolutionary systems for helping patients on Earth. Diseases behave and develop differently in microgravity, which assists scientists in finding cures.

Part B Translate the following paragraph into English.

聂海胜、刘伯明和汤洪波是中国天宫空间站的第一批居民，他们在空间站生活了90天。在此期间，他们进行了太空行走，使用大型机械臂（robotic arm）等设备在舱外安装和调试设备。他们还进行了多项科学实验，并与地球上的研究人员、教师和学生进行了视频通话。为期三个月的神舟十二号任务是中国第七次载人航天任务。

III Write a summary of the text in 120 words.

Directions: A summary should be written in your own words. It should contain only the ideas of the original text. Do not insert any of your own opinions, interpretations, deductions or comments into the summary.

IV Read the passage, select one word for each blank from a list of choices in the bank, and change the form when necessary. You may not use any of the words in the bank more than once.

| supply | crew | hazardous | follower | spacecraft |
| mission | agency | blast | account | launch |

A Russian humanoid robot was making its way on Thursday to the International Space Station after 1) _____ off on a two-week mission to support the crew and test its skills.

Known as FEDOR, which stands for Final Experimental Demonstration Object Research, the Skybot F-850 is the first humanoid robot to be sent into space by Russia. In 2011, NASA sent a humanoid robot, the Robonaut 2 into space to work in 2) _____ environments.

"The robot's main purpose is to be used in operations that are especially dangerous for humans onboard 3) _____ and in outer space," Russian space agency Roscosmos said on Thursday after the 4) _____ from the Baikonur Cosmodrome.

The ISS is a joint project of the space 5) _____ of the United States, Russia, Europe, Japan and Canada.

Traveling in an unmanned Soyuz MS-14 spacecraft, FEDOR is expected to dock at the ISS on Saturday with 660-kg cargo including medical 6) _____ and food rations for the 7) _____ working at the station, NASA said.

FEDOR, who is the size of an adult, has apparently embraced its 8) _____, describing itself as "an assistant to the ISS crew" on its Twitter page, which has 4,600 9) _____.

"Everything is normal," a tweet posted on its 10) _____ said a few hours after the flight.

V Critical thinking

Work in groups and discuss the following questions.
Almost every sector of human progress has benefited from sending people into space. However, it is argued that it is costly to send people into space, whose application is very limited in our daily life. Discuss whether or not space travel is worth the investment. Should we allocate the money invested in the space tech industry to other sectors?

Designs revealed for an incredible new space hotel

1 It has been more than 50 years since man first stepped on the moon, and we're still harboring dreams of escaping life on Earth for the mysteries of space. If a career as an astronaut isn't for you, perhaps the promise of a stay in a space hotel might be appealing.

2 Californian company the Gateway Foundation has released plans for the Von Braun Station, a cruise ship-style hotel floating among the stars. The aim is to get the hotel off the ground by 2025 and make it fully operational for travel by 2027.

3 The Von Braun Station is just one such space-based tourism option in development. Also planning to propel people into space are Virgin Galactic, Elon Musk's SpaceX company and Amazon CEO Jeff Bezos' Blue Origin aerospace company, not to mention the International Space Station – which announced the possibility of commercial collaborations. The Von Braun Station is also not the only space hotel design in the works. Earlier in 2019, US-based space tech startup Orion Span released plans for a luxury space hotel called Aurora Station, which it hopes to launch in 2022.

Among the stars

4 According to digitally rendered video and images released by the Gateway Foundation, the station resembles a rotating wheel, comprised of 24 modules, orbiting the Earth.

5 But how would the physics of the hotel work? Tim Alatorre, senior design architect at the Gateway Foundation, says the rotating wheel would create a simulated gravity. He further explains that "the station rotates, pushing the contents of the station out to the perimeter of the station, much in the way that you can spin a bucket of water – the water pushes out into the bucket and stays in place." Near the center of the station there's no artificial gravity, Alatorre says, but as you move down the outside of the

station, the feeling of gravity increases.

6 The Gateway Foundation's hotel design is named for Wernher von Braun, an aerospace engineer who pioneered rocket technology, first in Germany and later in the United States. The name was voted for by the Gateway Foundation members because the station is based on designs von Braun sketched out some 60 years ago.

7 "The basic physics of the station haven't changed since the 1950s, the way the station rotates," says Alatorre. The main difference is the modern materials – new metal alloys, carbon composites, 3D printing and launch pad technology that, says Alatorre, make a space hotel more probable in our current era.

8 Space tourism is an expensive game – Richard Branson's Virgin Galactic plans to launch passengers into sub-orbital space at the great sum of $250,000 per person, per trip. Meanwhile, Aurora Station says a stay in its space hotel will cost an eye-watering $9.5 million. Pricewise, in the early phases the Von Braun hotel will also be catering to those with dollars to spend, but the foundation is hoping to make it equivalent to "a trip on a cruise or a trip to Disneyland".

9 Aurora Station aims to sleep just 12, whereas the Von Braun Station will sleep 352 people with a maximum capacity of 450.

Warm aesthetics

10 So what will the Von Braun Station be like inside? Alatorre says the hotel's aesthetic was a direct response to the Stanley Kubrick movie *2001: A Space Odyssey*. "It was almost a blueprint of what not to do," says Alatorre. "I think the goal of Stanley Kubrick was to highlight the divide between technology and humanity and so, purposefully, he made the stations and the ships very boring and clean and alien."

11 Instead, Alatorre wanted to bring a slice of Earth to space, to avoid a laboratorial, overly Star Trek-esque feel. On board, there will be warm suites with carpets and stylish monochrome touches and chic bars that wouldn't look out of place back on Earth, just with star-gazing views.

12　There will also be plenty of fun recreational activities for guests to enjoy, says Alatorre. "We're going to have a number of different recreation activities and games that'll highlight the fact that you're able to do things that you can't do on Earth," he says. "Because of the weightlessness and the reduced gravity, you'll be able to jump higher, be able to lift things, be able to run in ways that you can't on Earth." A sport called "supersize basketball" is one such concept, according to Alatorre.

Starship culture

13　If it all sounds like a space-age trick, Alatorre is emphatic that the concept will have widespread, enduring appeal. "People will want to go and experience this just because it's a cool new thing and they've never done it before," he admits. "But our goal – the overall goal of the Gateway Foundation – is to create the starship culture where people are going to space, and living in space, and working in space and they want to be in space. And we believe that there's a demand for that." That means having space be a place where thousands of people are "living, working and thriving."

Sustainability in space

14　Given that the design is still just a design, there are some questions that remain unanswered about how the space hotel will function actually. For example, it's been suggested that living in low gravity for an extended period of time is damaging to the human body. While vacationers will probably only visit the hotel for a few weeks, staff will plan to be there for six months to a year.

15　They'll adjust schedules as needed, says Alatorre, but the foundation thinks this proposition would be perfectly safe. There's also the sustainability question, as people look for more eco-friendly vacations, surely going to space is not the solution?

16　Alatorre points to SpaceX's Raptor engine, which uses methane instead of petroleum-based fuel, suggesting "eco-friendly" rocket designs are the future. He says recycling will be woven into the fabric of the space hotel. "On the station itself, it's going to be about the most environmentally friendly vacation you'll ever have. Because we're recycling everything," says Alatorre. "There's no amount of water or trash or waste that is going to be discarded. Everything will be recycled, reused, stored, or converted to some other form."

New words

harbor /'hɑːbə/ *v.* to have in mind (a thought or feeling), usually over a long period 心怀

operational /ˌɒpə'reɪʃənəl◂/ *adj.* ready to be used 可使用的

startup /'stɑːtʌp/ *n.* a small business that has recently been started 新创办的小公司

sub-orbital /ˌsʌb'ɔːbɪtl/ *adj.* having a flight path that is less than one complete orbit of the Earth or another celestial body 亚轨道的

eye-watering /'aɪˌwɔːtərɪŋ/ *adj.* extremely surprising, because of being great in amount（在数目方面）令人极度惊讶的

laboratorial /ˌlæbərə'tɔːrɪəl/ *adj.* of a laboratory 实验室的

monochrome /'mɒnəkrəum/ *adj.* using different shades of one color 单色的

starship /'stɑːʃɪp/ *n.* a spacecraft designed for interstellar travel 星际飞船

sustainability /səˌsteɪnə'bɪlɨti/ *n.* the quality of being able to continue for a long time 可持续性

vacationer /və'keɪʃənə/ *n.* a holidaymaker 度假者

proposition /ˌprɒpə'zɪʃən/ *n.* a suggestion, or sth. that is suggested or considered as a possible thing to do 提议，建议

eco-friendly /'iːkəuˌfrendli/ *adj.* not harmful to the environment 环保的

methane /'miːθeɪn/ *n.* a gas without color or smell, that burns easily and is used as fuel 甲烷

recycle /ˌriː'saɪkəl/ *v.* to treat things that have already been used so that they can be used again 回收利用，再利用

Phrases and expressions

harbor dreams 心怀梦想
be comprised of 由…组成
the perimeter of 在…的周边
cater to 满足，迎合

highlight the divide between ... and ... 突出…与…之间的鸿沟
a slice of sth. …的一部分
Star Trek-esque feel 星际旅行式的感觉

Proper nouns

Gateway Foundation 美国加州盖特韦基
　金会

Von Braun Station 冯·布劳恩站太空
　酒店

Virgin Galactic 维珍银河公司

Blue Origin 蓝色起源公司

Orion Span 跨越猎户座公司

Aurora Station 极光站太空酒店

2001: A Space Odyssey《2001太空漫
　游》（美国科幻电影, 1968年上映）

Technical terms

digitally rendered video 数字渲染视频

simulated gravity 模拟重力

artificial gravity 人造重力

metal alloy 金属合金

carbon composite 碳复合材料

launch pad technology 发射台技术

sub-orbital space 亚轨道空间

Raptor engine 猛禽发动机

petroleum-based fuel 石油基燃料

Exercises

| For each of the following unfinished statements or questions, choose the most appropriate answer from A, B, C, or D according to the text.

1　According to the text, when can passengers visit the space hotel?

A. In 2025.

B. In the middle of the 21st century.

C. In the late 21st century.

D. In the early 21st century.

2　Walking to the outside of the station, visitors will find _____.

A. it difficult to breathe because of lack of gravity

B. it easy to stroll in outer space

C. it hard to adapt to the intense gravity

D. the sense of gravity becomes stronger

3   What will make the dream of building a space hotel come true?

   A. 3D printing and launch pad technology.

   B. New metal alloys.

   C. Carbon composites.

   D. All of the above.

4   The Von Braun hotel is designed for _____ according to the text.

   A. both average people and rich people

   B. celebrities

   C. experts in some field

   D. scientists

5   The decoration of the space hotel will be _____ that on Earth.

   A. similar to

   B. at odds with

   C. unrelated to

   D. absolutely different from

Questions for discussion

1   How much will it cost a family of five to travel to sub-orbital space?

2   What is the difference between the styles of the Stanley Kubrick movie *2001: A Space Odyssey* and Alatorre's design?

3   How is the term "starship culture" defined?

4   What are the problems to be settled for the space hotel?

5   What fuel will be likely used for the future rocket?

◉ **Practical writing**

✎ Language in academic writing

Academic writing delivers clear and accurate information, and attaches importance to well-constructed, carefully thought-out content. The use of proper language is essential in academic writing. English is also a common tool for publishing in international journals or participating in international conferences. This part will describe some aspects of language used in academic writing such as tense choice, verb voice, style, and sentence variety.

1. Tense choice

In academic writing the types of tenses which are commonly used by academic writers are simple present, simple past, and perfective aspects. Progressive aspects are rarely used in academic writing. Modality is also used, especially when the writers wish to make a recommendation or give an instruction. The future tense is used mainly in the discussion section of a research report and the work plan section of a research proposal.

2. Verb voice

The voice of the verb describes the relationship between the action expressed by the verb and the participants identified by its arguments. In academic writing, the passive voice is often preferred and the use of the first person, *I* or *we*, is generally avoided because using *I*, or even *we*, sounds egotistical. Avoiding them helps preserve the researcher's modesty. Most importantly, making *I* or *we* the subject of a sentence emphasizes the role of the researcher. In contrast, using the passive voice correctly emphasizes the research.

Examples:

1) We have investigated the influence of pH on the survival of E. coli bacteria.

2) The influence of pH on the survival of E. coli bacteria has been investigated.

3. Style

The language of academic writing should be formal. The language of academic English usually does not use language styles such as contractions (it didn't, they've, I won't, etc.) and hesitation fillers (er, um, well, you know, etc.). In academic writing the writer generally does not use personal pronouns, such as *I, you,* or *we.* Impersonal styles such as preparatory *it, there,* and *one* are often found in academic writing.

4. Sentence variety

Lengthy sentences make the text hard to read. Then again, too short sentences make for choppy writing without flow and cannot hold complex thoughts. With careful crafting, sentences are profound and readable. The following are some tips helping maintain an appropriate length of a sentence so that the message or idea can be communicated more effectively to the reader.

1) Varying the subject or word choice

One of the easiest ways to spot text that requires variety is by noting how each sentence opens. Writers often overuse the same word, like an author's name, or a subject, like pronouns to refer to an author, when beginning sentences. This lack of subject variety can be distracting to a reader. To avoid this type of repetition, try adjusting the placement of prepositional phrases or dependent clauses so the subject does not open each sentence.

2) Varying the sentence length

Another way to spot needed sentence variety is through the length of each sentence. Overusing longer sentences can overwhelm a reader and overshadow arguments, while frequently relying on shorter sentences can make one feel rushed or stunted. Alternating between lengths allows writers to use sentences strategically, emphasizing important points through short sentences and telling stories with longer ones.

Exercise

Modify the following sentences. Make sure to optimize the lengths of the sentences without losing their meanings.

1　In actual fact, every single nurse in the hospital worked for treating the patients from 3 a.m. in the morning to twelve midnight. (24 words)

2　Following instructions with a sentence too long can be confusing because it is easy to lose track of what was said at the beginning, since they do not give the reader enough time to process what they are reading and by the end of the sentence you might have forgotten where it started! (54 words)

3　The protein level was 10 mg in Group A, while it was 7 mg in Group B, the difference being statistically significant. (23 words)

4　The patient has iron-deficiency. His WBC count is also low. (11 words)

5　Although it was mandatory, the study participants were deprived of the right to sign their consent forms. (18 words)

6　A diagnosis of the cancer was made on the basis of the biopsy findings of the tumor. (18 words)

Unit 3
Aviation safety

Learning objectives

Upon completion of this unit, you will be able to:

- analyze possible factors causing plane accidents;
- find out lessons of plane accidents and precautions;
- understand the priority of safety that calls for joint efforts;
- write an abstract.

◉ Lead-in

Did a tired pilot incorrectly execute a procedure? Was an uneven landing performed due to inexperience? Aviation accidents continue to horrify people till this day, though safety has been the highest priority for the aviation industry over the past 100 years. According to the International Air Transport Association (IATA), future safety gains will come increasingly from analyzing data from millions of flights that operate safely every year. The industry's impressive safety record in recent decades is a reflection of the technological developments that have been introduced.

⊙ Intensive reading

Boeing faces safety questions after another 737 Max crash

1 The trouble appeared to begin almost immediately after takeoff. The pilots told air traffic controllers that they were having technical problems. The plane seemed to repeatedly climb and dive before a final plunge.

2 Two miserably similar scenes have played out in one year for Boeing 737 Max jets: on Mar. 10, 2019, when an Ethiopian Airlines flight crashed just after taking off from Addis Ababa, killing 157 people, and on Oct. 29, 2018, when a Lion Air disaster killed 189 people in Indonesia.

3 The Ethiopian crash occurred just outside the country's capital, leaving a smoking crater where investigators checked out the grim scene. Much about the cause of the crash remains unknown and will take weeks to investigate, and Boeing and the National Transportation Safety Board are sending teams to the crash site. But the rarity of two new planes of the same model going down in such a short time has urgently caught the attention of pilots, passengers, engineers and industry analysts.

4 For Boeing, the questions go to the heart of its business, as the 737 class is a workhorse for airlines worldwide, and the single-aisle 737 Max has been the company's best-selling plane ever. By the end of January 2019, Boeing had delivered over 350 737 Max jets since putting them in service in 2017. They have a list price of around $120 million, the company said, and around 5,000 more are on order.

5 "There's a whole lot of questions here and not a lot of answers," said John Cox, former executive air safety chairman of the Air Line Pilots Association in the United States.

6 The business of building and selling jets is cruelly competitive, and the 737 Max was Boeing's answer to an update that Airbus, the giant European aircraft manufacturer, displayed for its popular A320 jet that made it more fuel-efficient. The two companies are global leaders in the field, and they have competed for dominance for years. About

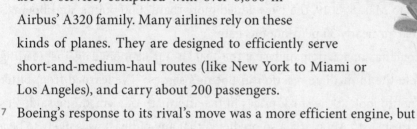

10,000 total planes from Boeing's 737 family are in service, compared with over 8,000 in Airbus' A320 family. Many airlines rely on these kinds of planes. They are designed to efficiently serve short-and-medium-haul routes (like New York to Miami or Los Angeles), and carry about 200 passengers.

7   Boeing's response to its rival's move was a more efficient engine, but the Max engine was bigger than the earlier versions. To address this engineering challenge, Boeing updated the software for the flight control system. After the Lion Air crash, some US aviation authorities said that the change had not been adequately explained to pilots.

8   But in light of the Indonesian disaster, pilots have since been informed by Boeing and regulatory agencies of the Max's new system and airlines have provided training classes on it. Whether Ethiopian Airlines, unlike Lion Air, which has a strong safety reputation, carried out that training was not immediately known.

9   On the Lion Air flight, the swings up and down may have come about as pilots repeatedly tried to keep that system from pushing the nose of the aircraft down, putting it into a fatal dive. Whatever happened, the Indonesian pilots lost their battle after about 12 minutes of flight.

10   While a malfunction of that system is a possibility in the Ethiopian flight, which lasted about six minutes and included a shorter series of swings, early information is still too general to draw conclusions. And what is known so far does not rule out pilot error or the malfunction of a completely separate system.

11   Robert Stengel, an expert on flight control systems and a professor of engineering and applied science at Princeton University, said it was not clear whether the rocking trajectory of the Ethiopian jet was caused by a malfunctioning control system or pilots trying to fly the plane manually while distracted by some other, as yet unknown, emergency. While cautioning that no conclusion could be reached without more information, Stengel said that broad similarities in the two crashes could not help but affect the flying public. "If you're simply looking at circumstantial evidence, that gives you pause, doesn't it?" Stengel said. "That's not a deep technical observation – that's just human nature."

12 Analysts agree that Wall Street is not going to be kind to Boeing stock. And whatever hit its shares take will weigh heavily on the Dow Jones Industrial Average, which in recent years has been lifted by Boeing's success. Shares of Boeing have tripled since the presidential election in 2016, making it the highest-priced stock in the Dow. From Nov. 8, 2016, to Mar. 8, 2019, the Dow added more than 7,000 points, and Boeing's rise accounted for nearly 30 percent of its gain.

13 Richard Aboulafia, an aviation analyst at the Teal Group, cautioned against reading too much into the immediate reaction in Boeing's shares. "I've learned from bitter experience not to look at the stock prices in the aftermath of a crash," he said. "It's just all over the place." Aboulafia also predicted that any pullback was likely to be a short drop, given the company's recent strength. At the close of trading on Mar. 8, 2019, Boeing was valued at nearly $239 billion, with a stock price above $422 a share. The company, which employs about 150,000 people, took in just over $100 billion in 2018, with profit for the year topping $10 billion.

14 Boeing has said that it is evaluating changes to the new software, and a person briefed on the matter said an upgrade was in the works. But after the Ethiopian crash, most Chinese carriers stopped using the dozens of Boeing 737 Max jets that they had acquired and began flying Boeing 737-800s on the same routes. The Civil Aviation Administration of China then issued an order for the required grounding of Boeing 737 Max planes by Chinese carriers.

15 China's main airlines are among the biggest users of the new Boeing 737 Max, having ordered at least 104 of them and taken delivery of at least 70. By contrast, many other carriers, often in slower-growing markets than China's, have taken delivery of only a small part of their orders for the Boeing 737 Max.

16 The pilots who fly the 737 Max are keeping a close eye on developments. Dennis Tajer, a spokesman for the American Airlines pilots union and a 737 pilot, said the training and other information had restored pilots' confidence in Boeing since the Indonesia crash. Tajer said it was too soon to say that the Ethiopian crash would change that view. But, he added, "We're not far off the time when we were being told that new equipment on the Max wasn't disclosed. We have that memory to work with." "It's too early," he said, to say what influence the crash will have, "but boy, are we watching."

New words

plunge /plʌndʒ/ *n.* a sudden movement
down or forwards 骤降；俯冲

crater /ˈkreɪtə/ *n.* a hole in the ground
made by something that has fallen on
it or by an explosion（由物体坠落、炸
弹爆炸形成的）坑

grim /grɪm/ *adj.* making you feel worried
or unhappy 令人担忧的；令人不愉
快的

rarity /ˈreərᵻti/ *n.* the quality of being rare
罕见，稀有

workhorse /ˈwɜːkhɔːs/ *n.* a machine or
person that you can rely on to do hard
and/or boring work 耐用的机器；吃苦耐
劳的人

single-aisle /ˌsɪŋgəlˈaɪl/ *adj.* having one
long passage between rows of seats
in a church, plane, theater, etc., or
between rows of shelves in a shop 单通
道的，单过道的

fuel-efficient /ˈfjuːəl ɪˌfɪʃənt/ *adj.*
burning fuel in a more effective way
than usual 节省燃料的

nose /nəʊz/ *n.* the pointed front part of a
plane, boat, etc.（飞机、船只等的）头部

malfunction /mælˈfʌŋkʃən/
n. the failure to function normally 故障
v. to fail to function normally 发生故障

caution /ˈkɔːʃən/ *v.* to warn sb. about a
problem, danger, etc. 警告

circumstantial /ˌsɜːkəmˈstænʃəl◂/ *adj.*
based on something that appears to be
true but is not proven 按情况推测的，
间接的

aftermath /ˈɑːftəmæθ/ *n.* the period
that follows an unpleasant event or
accident, and the effects that it causes
余波；后果

pullback /ˈpʊlbæk/ *n.* a reduction in the
value, amount or level of sth.（某物价
值、数量或水平的）降低

evaluate /ɪˈvæljueɪt/ *v.* to judge or
calculate the quality, importance, amount
or value of something 评估，评价

brief /briːf/ *v.* to give essential
information to sb. 向⋯介绍情况

grounding /ˈgraʊndɪŋ/ *n.* the process
of officially stopping an aircraft from
flying, especially because it is not safe
to fly（飞机的）留地停飞

Phrases and expressions

play out 发生; 出现

check out 调查, 核实

put ... in service 投入使用

compare with 与…比较

in light of 根据, 鉴于, 考虑到

come about 产生, 发生

can not help but do sth. 忍不住做某事

weigh on 重压于; 使烦恼

account for 占比; 解释

caution against 告诫不要做

all over the place 到处

value sth. at 给某物估价

brief sb. on 向某人提供信息

take delivery of 提取

by contrast 相比之下

keep a close eye on 密切关注

Proper nouns

Boeing 波音公司

Ethiopian Airlines 埃塞俄比亚航空公司

Addis Ababa 亚的斯亚贝巴 (埃塞俄比亚首都)

Lion Air 狮子航空公司 (简称 "狮航")

Air Line Pilots Association 美国民航飞行员协会

Airbus 空中客车公司 (简称 "空客")

Princeton University 普林斯顿大学

Wall Street 华尔街

American Airlines 美国航空公司 (简称 "美航")

Technical terms

air traffic controller 空中交通管制员

industry analyst 行业分析师

air safety 航空安全; 飞行安全

A320 jet A320系列飞机

short-and-medium-haul route 中短途路线

flight control system 飞行控制系统

regulatory agency 监管机构

Background information

Boeing 737 Max 波音737 Max机型

The Boeing 737 Max is the fourth generation of the Boeing 737, a narrow-body airliner manufactured by the American company Boeing. It is based on earlier 737 designs, with more efficient engines, aerodynamic changes and airframe modifications. The new series took its first flight in January 2016 and was certified by the US Federal Aviation Administration in March 2017.

National Transportation Safety Board 美国国家运输安全委员会
The National Transportation Safety Board (NTSB) is an independent federal agency charged by Congress with investigating every civil aviation accident in the United States and significant accidents in other modes of transportation – railroad, highway, marine and pipeline. It determines the probable cause of the accidents and issues safety recommendations aimed at preventing future accidents.

Dow Jones Industrial Average 道琼斯工业平均指数
The Dow Jones Industrial Average (DJIA), also known as Dow Jones, or simply the Dow, is a price-weighted measurement stock market index of 30 prominent companies listed on stock exchanges in the United States.

Civil Aviation Administration of China 中国民用航空局
The Civil Aviation Administration of China (CAAC) is the aviation authority under the Ministry of Transport of the People's Republic of China. It oversees civil aviation and investigates aviation accidents and incidents. As the aviation authority responsible for China, it concludes civil aviation agreements with other aviation authorities.

D ifficult sentences

1 **For Boeing, the questions go to the heart of its business, as the 737 class is a workhorse for airlines worldwide, and the single-aisle 737 Max has been the company's best-selling plane ever. (Para. 4)**

For Boeing, the questions are focused on the essential or core part of its business, because the 737 class is the most used plane for airlines in the world, and the single-aisle 737 Max has been the company's most popular plane on the market.

2 **The business of building and selling jets is cruelly competitive, and the 737 Max was Boeing's answer to an update that Airbus, the giant European aircraft manufacturer, displayed for its popular A320 jet that made it more fuel-efficient. (Para. 6)**

It is very competitive to build and sell jets. The 737 Max was manufactured by Boeing in response to Airbus, the European aircraft leader. Its A320 jets are known for their consumption of less fuel.

3   **While a malfunction of that system is a possibility in the Ethiopian flight, which lasted about six minutes and included a shorter series of swings, early information is still too general to draw conclusions. (Para. 10)**

A failure of the flight control system might result in the crash of the Ethiopian flight. The plane flew for six minutes with short climbing and descending. But it is too early to come to any conclusions due to a lack of detailed information.

4   **Robert Stengel, an expert on flight control systems and a professor of engineering and applied science at Princeton University, said it was not clear whether the rocking trajectory of the Ethiopian jet was caused by a malfunctioning control system or pilots trying to fly the plane manually while distracted by some other, as yet unknown, emergency. (Para. 11)**

Robert Stengel is an expert on flight control systems and a professor of engineering and applied science at Princeton University. He said the reason of the plane's unstable condition was unclear. It may be either breakdown of the control system, or pilots who tried to control the plane by hand but failed because of some unknown emergency.

5   **While cautioning that no conclusion could be reached without more information, Stengel said that broad similarities in the two crashes could not help but affect the flying public. (Para. 11)**

Stengel warned that we could not reach a conclusion about how the accidents happened since we don't have enough information. As there were a lot of similarities between the two accidents, airline passengers have reason to be worried.

6   **Analysts agree that Wall Street is not going to be kind to Boeing stock. And whatever hit its shares take will weigh heavily on the Dow Jones Industrial Average, which in recent years has been lifted by Boeing's success. (Para. 12)**

Analysts hold that Wall Street will not treat Boeing stock kindly. And whatever blow its shares suffer will put great pressure on the Dow Jones Industrial Average, although it has been encouraged by Boeing's success in recent years.

7   **Richard Aboulafia, an aviation analyst at the Teal Group, cautioned against reading too much into the immediate reaction in Boeing's shares. (Para. 13)**

Richard Aboulafia is an aviation expert at the Teal Group. He warned people not to be too sensitive about the recent response to Boeing's shares.

8   **Aboulafia also predicted that any pullback was likely to be a short drop, given the company's recent strength. (Para. 13)**

In addition, Aboulafia indicated that Boeing's fall in share prices may not last long, considering the company's recent growth.

9   **Boeing has said that it is evaluating changes to the new software, and a person briefed on the matter said an upgrade was in the works. (Para. 14)**

Boeing has announced that it is assessing changes to the new software. A person who was informed about the matter said that a software improvement was being made.

10 **But, he added, "We're not far off the time when we were being told that new equipment on the Max wasn't disclosed. We have that memory to work with." (Para. 16)**

But, he added, "We were told not long ago that details on the new equipment were not given. We still remember it clearly."

Exercises

| **Fill in the blanks with the words and expressions given below, and change the forms when necessary.**

| compare with | evaluate | account for | triple | caution against |
| grounding | value at | keep a close eye on | aftermath | malfunction |

1   In the _____ of the tragic event there is a growing recognition that poverty is a cause of violent conflict.

2   Due to the economic crisis, the house was _____ less than the price the seller was asking.

3   If such a weapon is being designed with that specific purpose, how do you _____ the risk?

4   Jack must be sick; it's the only thing that will _____ his strange behavior.

5   _____ the traditional way of shopping, we have no need to go out to pick up our favorite goods from store to store.

6   The workers have _____ in number since the company opened two years ago.

7   All parties stress diplomacy and sanctions to dissuade the country from developing nuclear weapons, and _____ a rush to war.

8   Christina, please _____ this pot so that the soup doesn't boil over.

9   Commercial carriers at the airport started flying for the first time since a storm forced the _____ of almost all planes.

10   China's first lunar rover, Yutu, or Jade Rabbit, is still "alive" despite the _____ of some equipment, according to a senior space scientist.

‖ Translation

Part A   Translate the following paragraph into Chinese.

The business of building and selling jets is cruelly competitive, and the 737 Max was Boeing's answer to an update that Airbus, the giant European aircraft manufacturer, displayed for its popular A320 jet that made it more fuel-efficient. The two companies are global leaders in the field, and they have competed for dominance for years. About 10,000 total planes from Boeing's 737 family are in service, compared with over 8,000 in Airbus' A320 family. Many airlines rely on these kinds of planes. They are designed to efficiently serve short-and-medium-haul routes (like New York to Miami or Los Angeles), and carry about 200 passengers.

Part B   Translate the following paragraph into English.

人们需要制定规则来确保无人机的安全性，如今许多国家都已制定此类法规。总体来说，因为不想被吊销飞行许可证，或被没收装备，专业操作手和热心的业余爱好者会遵守这些法规。严厉的惩罚措施和更好的信息沟通可以管控不负责任的用户。制造商还可在无人机的数字导航系统中设置安全装置，以防无人机飞得太高或离机场等敏感区域太近。

‖‖ Write a summary of the text in 120 words.

Directions: A summary should be written in your own words. It should contain only the ideas of the original text. Do not insert any of your own opinions, interpretations, deductions or comments into the summary.

‖V Read the passage, select one word for each blank from a list of choices in the bank, and change the form when necessary. You may not use any of the words in the bank more than once.

| prohibit | information | belong | allow | acceptable |
| cabin | material | screening | eliminate | requirement |

One of the relatively newer regulations is the 1) _____ for all passengers to present an acceptable ID at airport checkpoints. For air travel to or within the US, the Transportation Security Administration (TSA) requires adult passengers (18 and over) to carry a US federal or state-issued photo ID with the following 2) _____: name, date of birth, gender, and expiration date.

If a(n) 3) _____ ID is not provided, the passenger will have to prove their identity to a transportation security officer, and go through further 4) _____. Individuals who cannot prove their identity are not allowed to fly.

The regulation allows for typical, everyday liquids in small quantities to be carried on the airplane. This helps 5) _____ the danger of having explosive liquids and materials onboard. The TSA also maintains a list of prohibited items, with distinctions being made between carry-on and checked items. Sharp, dangerous items are not allowed in the 6) _____, including knives, box cutters, and razor blades. However, these items are 7) _____ onboard as checked items. Other item categories that follow these guidelines are sporting goods, tools, self-defense items and firearms.

Any item 8) _____ to the explosive and flammable 9) _____, disabling chemicals, and other dangerous item categories must be completely 10) _____ and not allowed onboard the airplane.

V Critical thinking

Work in groups and discuss the following questions.
The aviation's success in safety is collectively attributed to carefully reviewing past experiences, evaluating in detail how things went wrong, and making the necessary and timely changes to prevent the same accidents from happening again. In addition, advances in technology also contribute to safety in the aviation industry. However, there are still reports of air accidents from time to time. What are the possible reasons for those accidents? Do you think AI technology can be used to improve the levels of aviation safety? Give details to support your view.

◉ **Further reading**

Unmanned aircraft safety, risk management and insurance

1　Drones, also known as unmanned aerial vehicles (UAVs), or components of an unmanned aerial system (UAS), are usually associated with the military and as toys for grown-ups, yet it seems that the works are already underway to create UAV aircraft that transport goods, materials, and even people. Amazon, the largest online retailer in the world, has already placed a patent on a flying warehouse which is predicted to revolutionize the freight industry as a whole. Once this takes effect, it is only a matter of time before commercial UAVs carrying passengers will begin production in earnest.

2　In February of 2019, the Federal Aviation Administration (FAA) released its Notice of Proposed Rulemaking (NPRM) regarding UAVs. It appears to have been an important moment for corporations that had previously adopted a wait-and-see approach about investing further in UAV manufacturing and operations. At Global Aerospace, we have observed an increase in inquiries regarding insurance for commercial UAV operations.

3　As applications for UAVs increase, so do concerns about privacy and safety. As a result, managing and insuring against risk will be crucial to success for UAV manufacturers and operators.

4　From an insurance perspective, the proliferation of the small commercial UAVs covered by the NPRM presents a new set of challenges, not least of which is the lack of available industry data. While commercial airline and general aviation accidents are hard to predict using even the most sophisticated modeling tools, insurers at least have a good sense of the premium they need to charge to cover the likely loss activity in any given year.

5 However, with commercial UAVs, there is little data upon which to make similar predictions. Most models of UAVs have not existed long enough for insurers to acquire an understanding of the particular features that could influence the likelihood of an accident or system failure. Another hurdle to handle is the wide range of experience that UAV operators (pilots) have when they start in the UAV business.

Training

6 One primary risk management tool for UAVs that insurers will be looking at is training. Without effective training in the hazards involved, UAV operators will never be able to operate at optimal safety. The NPRM indicated that operators would have to pass an aeronautical knowledge test. It is likely that this will include the need to demonstrate an understanding of aeronautical charts, meteorology, aerodynamics and more.

7 Training for all levels of UAV operation is becoming widely available, from an online course for approximately $200 offered by organizations including the Unmanned Safety Institute, to custom training for a corporation's team of operators. Some insurance providers already require operators to undertake some type of formal training.

8   Another issue related to safety training is the quality of the operating manual and after-sales support. Important information, such as the relative battery deterioration in cold weather, is missing from many instruction manuals.

9   At Global Aerospace, we applaud the manufacturers who are taking a proactive position on safety training. It is our perspective that the UAV manufacturers who take an integrated approach to both sales and safety will achieve the greatest long-term success.

Safety management for UAVs

10   Safety documents such as pre-flight checklists, logbooks and a Standard Operating Procedure (SOP) are established components of manned aviation at all levels. These documents come under the general heading of a Safety Management System (SMS). At Global Aerospace, the belief in the importance of SMS documents as a risk reduction tool led us to arrange for a SOP manual from the Unmanned Safety Institute. We are making this manual available to all Global Aerospace UAV customers.

Maintenance

11   The FAA's NPRM outlined the need for UAVs to be maintained in a suitable condition for safe operations. Comment was invited on the maintenance and inspection proposals, and we may yet see a requirement for required periodic inspections by approved facilities. Regardless of how the final rules are drafted, the responsibility will fall on the operator to ensure the UAV is inspected prior to each flight and is in a suitable condition for safe operation.

Environmental hazards

12   The NPRM outlined the need for UAVs to avoid flying over people not directly involved in the operation. The failure rate of small UAVs is still too high to take such chances where the risk of serious injury exists. Many start-up technology companies are working on solutions for these UAV-related risks. Geo-fencing, or the ability to build technology into the software to prevent a UAV from flying in restricted airspace, is already available and in use. More manufacturers are integrating it into their products.

Privacy issues

13   The final key UAV risk factor to consider is respecting people's privacy. Using UAVs in a responsible and ethical manner will ultimately lead to a lower risk profile as well as greater public acceptance of this new technology. Simple precautions can be taken

to avoid breaching an individual's reasonable expectation of privacy. These could include gaining the person's consent to being filmed and taking care not to publish any images or material captured without their consent.

Insurance for unmanned aviation

14 As with any aerospace operation, insurance is an integral part of risk management. It is there to provide financial compensation when the safety management system has failed to prevent an accident or a loss has been suffered due to an unforeseen event. While the regulatory situation continues to evolve and change, the subject of insurance is increasingly important within the UAV community. Owners and operators, as well as manufacturers and other service providers, are all interested in insurability and the cost of premiums. Many in the UAV industry are looking to insurance carriers to be the driving force of the various risk management initiatives currently in development. Common questions that come up regarding UAV insurance include: Do I need insurance for my UAV? How much does UAV insurance cost? Do I need to be approved by the FAA to obtain UAV insurance? What would commercial UAV insurance cover?

15 While the FAA's NPRM made no mention of insurance requirements, any commercial UAV operator should assume that their customers and partners will eventually require them to certify that they are insured. In any event, we anticipate that most professional UAV operators will purchase insurance for legal liability and to protect their assets.

The insurance market for UAVs

16 Aviation insurance carriers in the US that are active in the UAV sector offer different solutions and levels of coverage. Some, such as Global Aerospace, have UAV-specific policies and coverage. The insurability of an operation depends upon a number of factors including: choice of platform, experience of the operators, and intended use.

17 Perhaps most of all, insurers assess the likelihood of an accident involving people, as that is where the possibility of expensive litigation and indemnity payments exists.

18 If liability limits higher than a few million dollars are required, the insurance marketplace is reduced to just a handful of available carriers. The higher the limit, the more questions about safety and operating procedures will be asked. Insurers routinely demand higher safety standards than those set by the FAA for traditional aviation risks.

New words

aerial /ˈeəriəl/ *adj.* in or moving through the air 空中的；空气中的

revolutionize /ˌrevəˈluːʃənaɪz/ *v.* to completely change the way that sth. is done 彻底改变；完全变革

operator /ˈɒpəreɪtə/ *n.* a person who operates equipment or a machine 操作员

proliferation /prəˌlɪfəˈreɪʃən/ *n.* a sudden increase in the number or amount of sth. 激增

premium /ˈpriːmiəm/ *n.* an amount of money that you pay once or regularly for an insurance policy 保险费

aeronautical /ˌeərəˈnɔːtɪkəl/ *adj.* involving or relating to the science or practice of building or flying aircraft 航空（学）的

meteorology /ˌmiːtiəˈrɒlədʒi/ *n.* the study of the processes in the Earth's atmosphere that cause particular weather conditions, especially in order to predict the weather 气象学

aerodynamics /ˌeərəʊdaɪˈnæmɪks/ *n.* the study of the way in which objects move through the air 空气动力学

deterioration /dɪˌtɪəriəˈreɪʃən/ *n.* the process of changing to an inferior state 恶化

pre-flight /priːˈflaɪt/ *adj.* of or relating to the period just prior to the flight of a plane 飞机起飞前的

checklist /ˈtʃekˌlɪst/ *n.* a list of the things that you must remember to do, to take with you or to find out（记事）清单，一览表

logbook /ˈlɒgbʊk/ *n.* an official record of events during the journey in a plane or ship 飞行日志；航海日志

periodic /ˌpɪəriˈɒdɪk◂/ *adj.* happening fairly often and regularly 周期性的；定期的

compensation /ˌkɒmpənˈseɪʃən/ *n.* sth., especially money, that sb. gives you because they have hurt you, or damaged sth. that you own; the act of giving this to sb. 补偿（或赔偿）物；（尤指）补偿金，赔偿金；赔偿

certify /ˈsɜːtɪ̩faɪ/ *v.* to state officially, especially in writing, that sth. is true（尤指书面）证明，证实

liability /ˌlaɪəˈbɪl̩ti/ *n.* the state of being legally responsible for sth.（法律上对某事物的）责任，义务

litigation /ˌlɪtɪ̩ˈgeɪʃən/ *n.* the process of making or defending a claim in court 诉讼

indemnity /ɪnˈdemn̩ti/ *n.* a sum of money that is given as payment for damage or loss 赔款；补偿金

Phrases and expressions

place a patent on 对…申请专利
adopt a wait-and-see approach 采取观
　望的态度
start-up technology company 初创技术

公司
make mention of 提及
liability limit 责任限额

Proper nouns

Amazon 亚马逊（全球著名网络零售商）
Federal Aviation Administration (FAA) 美
　国联邦航空管理局
Notice of Proposed Rulemaking (NPRM)
　《建议制定规则的通知》
Global Aerospace 环球宇航公司

Unmanned Safety Institute 无人机安
　全学院
Standard Operating Procedure (SOP)
　标准操作程序
Safety Management System (SMS) 安
　全管理系统

Technical terms

unmanned aerial vehicle (UAV) 无人机
unmanned aerial system (UAS) 无人机
　系统
general aviation 通用航空
modeling tool 建模工具
aeronautical chart 航空地图（简称

"航图"）
instruction manual 指导手册
pre-flight checklist 飞行前检查表
manned aviation 载人航空
geo-fencing 地理围栏；区域限定
restricted airspace 空中禁区

Exercises

I For each of the following unfinished statements or questions, choose the most appropriate answer from A, B, C, or D according to the text.

1　What will be necessary to achieve success for those who manufacture and operate UAVs?

　　A. Manufacturing and selling UAVs.

　　B. Managing and insuring against risk.

　　C. Maintaining UAVs.

　　D. Publicizing UAVs.

2　Using drones in a responsible way will bring about the following results EXCEPT _____.

　　A. a lower risk profile

　　B. greater public acceptance of the new technology

　　C. avoiding breaching an individual's reasonable expectation of privacy

　　D. publishing any images or material captured

3　Owners, operators, manufacturers, and other service providers are all concerned about _____.

　　A. after-sales service of drones

　　B. market prospects of drones

　　C. insurability and the cost of premiums

　　D. maintenance of drones

4　Whether an operation can be insured depends upon a number of factors EXCEPT _____.

　　A. flight time

　　B. choice of platform

　　C. experience of the operators

　　D. intended use

5　Who generally call for stricter safety standards than those set by the FAA for traditional aviation risks?

　　A. Pilots.

　　B. Owners.

　　C. Manufacturers.

　　D. Insurers.

|| Questions for discussion

1 What has been found after the FAA released its NPRM on unmanned aerial vehicles?

2 What are people worried about with the increase of applications for unmanned aerial vehicles?

3 What should operators understand in order to pass an aeronautical knowledge test?

4 What are the operators expected to ensure no matter how the final rules are drafted?

5 Why is insurance critically important for any aerospace operation?

◉ Practical writing

How to write an abstract

An abstract is a shortened form of a paper, article, book, etc., giving only the most important facts or arguments in a condensed form. It acts as a window through which the writer illustrates his/her research orientation or discussion so that readers can decide whether the writing is worthy of reading or not. Abstracts help readers decide whether the writing is relevant, useful, or not.

Basically, there are two types of abstracts: informative and descriptive. Informative abstracts outline the scope and content of a paper, including research purposes, methods, results and conclusions. They are used for much longer and technical research papers and are particularly suitable for experimental studies. Descriptive abstracts mainly introduce the point of view or research purposes of the paper rather than the research methods, findings and conclusions. In addition, no explanation or comment should be added. They are best for shorter papers. Abstracts are typically a paragraph or two, and should be no more than 10% of the length of the full essay.

The readers for abstracts can be broad – from experts to lay people. You should write an abstract that both provides technical information and remains comprehensible to non-experts. Keep technical language to a minimum, as the readers may not have the same level of knowledge as you do.

Keywords appear at the end of an abstract. They are particular words or phrases that describe the content of a paper, thesis, report, etc. Keywords can be entered into a search engine in order to find webpage results that match the user's purpose.

Here are some tips for writing an abstract:
1) Don't make the abstract too general or too specific.

2) Avoid simply copying the introduction or first paragraph as an abstract.

3) Pay attention to verb tense and aspect.

4) Be sure that the keywords come up in the abstract.

5) Finally, make sure that the abstract covers the reasons for the studies, a problem statement, methods of solving problems, results and conclusions.

Exercise

Find an informative abstract and a descriptive abstract online. Check how they are different, and whether the guidelines mentioned above are observed in them.

Unit 4
Women pilots in aviation

Learning objectives

Upon completion of this unit, you will be able to:

- identify the leading female pilots and their influences in the industry;
- clarify how women are involved in aviation in the world;
- evaluate the realization of gender equality in the history of aviation;
- illustrate data by using figures and tables.

 Lead-in

In contemporary society, gender equality is still an issue to be resolved in the fierce competition of the global economy. Women pilots, as a new-born profession, have caught the eye of the world. On the one hand, as women pilots have entered the workforce of aviation and military service, what have they done to pursue gender equality in the process? On the other hand, though with wings, female pilots have made history over and over again, why still are so few women choosing to work in aviation?

⊙ **Intensive reading**

American women involved in aviation

1 Today, women pilots fly for the airlines, fly in the military and in space, fly air races, command helicopter mercy flights, haul freight, seed clouds, patrol pipelines, teach students to fly, maintain jet engines, and transport corporate officers.

2 Women have made a significant contribution to aviation since the Wright brothers' first 12-second flight in 1903. Blanche Scott was the first American woman pilot, in 1910, when the plane that she was allowed to taxi mysteriously became airborne. In 1911, Harriet Quimby became the first licensed woman pilot. And in 1912, Harriet became the first woman to fly across the English Channel.

3 In 1921, Bessie Coleman became the first African-American woman pilot. Because of the discrimination in the United States toward women as pilots and her race, Bessie moved to France and learned to fly at the most famous flight school in France. She returned to the United States and pursued a barnstorming career until 1926.

4 On March 16, 1929, Louise Thaden made her bid for the women's endurance record in a Travel Air, from Oakland Municipal Airport, California, and succeeded with a flight of 22 hours, 3 minutes. The record was broken a month later by Elinor Smith with 26 hours, 21 minutes over Roosevelt Field, New York.

5 Other firsts followed: Katherine Cheung, in 1932 in Los Angeles, California, was the first woman of Chinese ancestry to earn a license. Anne Morrow Lindbergh, wife of the famous pilot Charles Lindbergh, was the first US woman glider pilot and first woman recipient of the National Geographic Society's Hubbard Award. And, Phoebe Fairgrave Omlie was the first woman transport pilot. Phoebe, considered to be one of America's top women pilots in the 1920s and 1930s, developed a program for training women flight instructors and was appointed as Special Assistant for Air Intelligence of the National Advisory Committee for Aeronautics (the forerunner of NASA), and was active in the National Air Marking Program to paint airport identification symbols on airports or nearby buildings.

6   Air racing was a way for women to demonstrate their abilities, and of course, the prize money was an incentive. All-women's air races were soon organized, the biggest being the National Women's Air Derby in 1929. The race was from Santa Monica, California to Cleveland, Ohio and flown in eight days. The idea of letting women race airplanes was not accepted by many people. During the air race there were threats of sabotage and headlines that read "Race Should Be Stopped." However, the race drew twenty women from across the country and gave them the chance to meet face-to-face for the first time.

7   After the race, these women kept in contact with each other and talked about forming an organization of women pilots. Clara Trenckman, who worked in the Women's Department of the Curtiss Flying Service at Valley Stream, Long Island, convinced two Curtiss executives to invite licensed women to meet in Valley Stream to form such an organization. Responding to the invitation, 26 licensed women pilots met in a hangar at Curtiss Field on November 2, 1929 to formally create The 99s Club. Later, after many rejected names, the organization chose its name "The Ninety-Nines" because 99 of the 117 licensed American women pilots in the United States at that time signed up as chartered members.

8   By 1930 there were about 200 women pilots, and by 1935 there were between 700 and 800 licensed women pilots. A major breakthrough in aviation was allowing women to air race against men. In 1936, Louise Thaden and Blanche Noyes won the prestigious Bendix Trophy race. Women have competed against men ever since.

9   As World War II progressed, women were able to break into many aspects of the aviation world. They served as test pilots, mechanics, flight controllers, instructors, and aircraft production line workers. At the beginning of 1943, 31.3 percent of the aviation workforce were women. World War II pushed the movement of women into aviation fields. The history of aviation during these years is immense.

10 The Women's Auxiliary Ferry Squadron (WAFS), founded by Nancy Harkness Love, and the Women's Flying Training Detachment (WFTD), founded by Jacqueline Cochran, were fused together by President Roosevelt to become the Women Airforce Service Pilots (WASP). The new organization was a vital part of the history of women in military aviation. Although these women were civilians and outnumbered by

women in the regular military service of World War II, their experiences present a paradigm for the service of WWII military women. Unfortunately, the WASP members were not recognized as military personnel until the US Senate passed a resolution in November 1977 and it was signed into law by President Carter.

11 The years since World War II have brought down many more barriers for women pilots and records continue to be broken. Jacqueline Cochran went on to be the first woman pilot to break the sound barrier, with Chuck Yeager acting as her chase pilot, on May 20, 1953. In 1954 Marion Hart flew the Atlantic at the age of 62.

12 Women got their first step closer to space in 1959, when Geraldyn Cobb, a talented young pilot, became the first woman to undergo the Mercury astronaut physiological tests. She was 28 years old, had 7,000 hours of flight time, and held three world records. Cobb successfully completed all three stages of the physical and psychological tests that were used to select the original seven Mercury astronauts. Although 13 women finished this first round of testing, NASA refused to authorize the completion of the tests for fear that such action might be taken as approval of female astronauts.

13 Not even the Soviet Union's launch of Valentina Tereshkova into space in 1963, nor the Civil Rights Act of 1964 broke ground for women in space. It was not until June 17, 1983, that Dr. Sally Kristen Ride, NASA astronaut and a South Central Section 99, made history as the first US woman in space, serving as a specialist for STS-7 on the six-day flight of the Space Shuttle orbiter Challenger.

14 By the 1960s there were 12,400 licensed women pilots in the United States (3.6 percent of all pilots). This number doubled by the end of the decade to nearly 30,000 women, but was still only 4.3 percent of the total pilots. Today, women comprise about 6 percent of pilots in the United States.

15 People become pilots for the same reasons. First, they love flying; second, they love using their talents and being respected for them; and mostly, they love the feeling of belonging to this strong family called aviation.

New words

barnstorming /ˈbɑːnˌstɔːmɪŋ/ *adj.* done with a lot of energy and very exciting to watch 精彩的, 令人兴奋的

ancestry /ˈænsəstri/ *n.* the members of your family who lived a long time ago 祖先

glider /ˈglaɪdə/ *n.* an aircraft without an engine, which flies by floating on air currents 滑翔机

recipient /rɪˈsɪpiənt/ *n.* the person who receives sth. 接受者

incentive /ɪnˈsentɪv/ *n.* something that encourages you to work harder, start a new activity, etc. 鼓励, 激励; 动力; 刺激

sabotage /ˈsæbətɑːʒ/ *n.* deliberate damage, for example, in a war or as a protest 蓄意破坏

hangar /ˈhæŋə/ *n.* a large building in which aircraft are kept 飞机库; 飞机棚

outnumber /aʊtˈnʌmbə/ *v.* to have more people or things in number than the other group 数目超过; 比⋯多

paradigm /ˈpærədaɪm/ *n.* a model of something, or a very clear and typical example of something 示例, 样式

resolution /ˌrezəˈluːʃən/ *n.* a formal decision made at a meeting by means of a vote 决议

Mercury /ˈmɜːkjʊ̈ri/ *n.* the planet closest in distance to the Sun 水星

physiological /ˌfɪziəˈlɒdʒɪkəl/ *adj.* of or relating to the biological study of physiology 生理的; 生理学的

authorize /ˈɔːθəraɪz/ *v.* to give official permission for something 批准; 认可

Phrases and expressions

make a bid for 设法得到; 出价购买
break a record 打破纪录
sign up as 注册为

be fused together by 由⋯融合成
for fear that 生怕, 唯恐
break ground 开辟新天地; 破土动工

Proper nouns

English Channel 英吉利海峡

National Geographic Society 美国国家地理学会

Hubbard Award 哈伯德奖

Curtiss Flying Service 柯蒂斯飞行服务队

Women's Auxiliary Ferry Squadron (WAFS) 妇女辅助渡运中队

Women's Flying Training Detachment (WFTD) 妇女飞行训练支队

President Roosevelt 罗斯福总统（美国第32任总统）

Women Airforce Service Pilots (WASP) 女子航空勤务飞行队

US Senate 美国参议院

President Carter 卡特总统（美国第39任总统）

Technical terms

air race 飞行竞赛

mercy flight 救援飞行

haul freight 运输货物

jet engine 喷气发动机

test pilot 试飞员

flight controller 飞行指挥员

chase pilot 护航飞行员

Background information

National Advisory Committee for Aeronautics 美国国家航空咨询委员会

The National Advisory Committee for Aeronautics (NACA), the forerunner of NASA, was a federal agency founded on March 3, 1915, for promoting and conducting aeronautical research. Over the years, NACA made fundamental aeronautical breakthroughs in the way today's aircraft and spacecraft are built, tested and designed. As the NACA moved toward the edge of space, on October 1, 1958, its assets and personnel were transferred to the newly created National Aeronautics and Space Administration (NASA).

National Women's Air Derby 美国国家女子飞行竞赛

The National Women's Air Derby was the first official women-only air race in the United States, taking place during the 1929 National Air Races.

Civil Rights Act of 1964《1964年民权法案》

The Civil Rights Act of 1964 is a landmark civil rights and labor law in the United States that outlaws discrimination based on race, color, religion, sex, national origin, and later gender identity. The act remains one of the most significant legislative achievements in American history.

Difficult sentences

1   **Today, women pilots fly for the airlines, fly in the military and in space, fly air races, command helicopter mercy flights, haul freight, seed clouds, patrol pipelines, teach students to fly, maintain jet engines, and transport corporate officers. (Para. 1)**

Here, it refers to diverse jobs for women pilots, flying for different organizations and purposes, such as flying to rescue people, to deliver goods, to make artificial rain, to check pipelines, to teach students to fly, to keep jet engines in a good condition, and to transport officers. It serves as the topic sentence of the whole passage, indicating the position of current women pilots with a comparison of their position in the past shown in the following paragraphs.

2   **Blanche Scott was the first American woman pilot, in 1910, when the plane that she was allowed to taxi mysteriously became airborne. (Para. 2)**

Blanche Scott was training in a plane and it flew into the air in 1910, without official approval. This made her the first American woman pilot in history.

3   **On March 16, 1929, Louise Thaden made her bid for the women's endurance record in a Travel Air, from Oakland Municipal Airport, California, and succeeded with a flight of 22 hours, 3 minutes. (Para. 4)**

On March 16, 1929, Louise Thaden made great efforts to fly from Oakland Municipal Airport, California, in a Travel Air plane. It became a record for the women's endurance flying – 22 hours, 3 minutes, without landing.

4   **World War II pushed the movement of women into aviation fields. The history of aviation during these years is immense. (Para. 9)**

For the demand of military service in WWII, a majority of men had been recruited and sent to the whole world. The civil service was faced with the shortage of labor, and that's why the workforce of women played an important role during the war. The movement of women into aviation fields also benefited from the situation.

5   **Unfortunately, the WASP members were not recognized as military personnel until the US Senate passed a resolution in November 1977 and it was signed into law by President Carter. (Para. 10)**

The WASP members were not considered part of military personnel for a long time. It was in November 1977 that the US Senate passed a resolution that acknowledged officially their military status, and President Carter signed it into law.

6   **Although 13 women finished this first round of testing, NASA refused to authorize the completion of the tests for fear that such action might be taken as approval of female astronauts. (Para. 12)**

The part "for fear that …" is used in the subjunctive mood with a structure of "might + *v*." in the adverbial clause. Here, it means that NASA was unwilling to approve the status of female astronauts by refusing to acknowledge the completion of the tests.

7   **Not even the Soviet Union's launch of Valentina Tereshkova into space in 1963, nor the Civil Rights Act of 1964 broke ground for women in space. (Para. 13)**

The phrase "not ... nor ..." means "neither ... nor ..." or "both ... and ... not". Valentina Tereshkova was launched into space in 1963 by the Soviet Union, and the Civil Rights Act was issued in 1964. However, both these two significant events did not establish the identity for women in space.

8   **It was not until June 17, 1983, that Dr. Sally Kristen Ride, NASA astronaut and a South Central Section 99, made history as the first US woman in space, serving as a specialist for STS-7 on the six-day flight of the Space Shuttle orbiter Challenger. (Para. 13)**

The sentence is in the emphatic pattern to emphasize the date June 17, 1983, when Dr. Sally Kristen Ride set her first step in space. She was a NASA astronaut, a member of The Ninety-Nines, and a specialist for STS-7 on the six-day flight of the orbiter Challenger.

Exercises

I Fill in the blanks with the words and expressions given below, and change the forms when necessary.

> make a bid for   outnumber   sign up as   authorize   recipient
> resolution   sabotage   incentive   ancestry   for fear that

1 By then, most biologists had come to agree with Darwin that species shared a common _____.

2 I felt glad that I was now a giver of pleasure, not merely a passive _____.

3 Economic returns from these projects may provide a strong _____ for such countries to participate.

4 Officials have not yet ruled out _____ as a possible cause of the crash.

5 The government has announced recently that it will build more non-profit public hospitals to ensure they _____ privately-owned hospitals and patients get better medical services.

6 The United Nations passed a(n) _____ to increase aid to developing nations.

7 They were both _____ to conduct political negotiations as far as necessary to coordinate the war effort.

8 The young salesman is working hard to _____ promotion to a regional sales manager next year.

9 People can find out about the service, _____ members, find and reserve nearby cars, and manage their accounts.

10 It is as if we believed it best to direct the consumer to a few trusted companies, _____ he might choose to deal with a young unpolished start-up.

II Translation

Part A   Translate the following paragraph into Chinese.

By the 1960s there were 12,400 licensed women pilots in the United States (3.6 percent of all pilots). This number doubled by the end of the decade to nearly 30,000 women, but was still only 4.3 percent of the total pilots. Today, women comprise about 6 percent of pilots in the United States.

People become pilots for the same reasons. First, they love flying; second, they love using their talents and being respected for them; and mostly, they love the feeling of belonging to this strong family called aviation.

Part B   Translate the following paragraph into English.

在美国、英国等一些发达国家，虽然在过去几十年也有女性参加民航培训，但实际上女飞行员的比例仍然很低。据统计，目前中国、美国、英国等16个国家的军队中都有女飞行员。其中，中国是有女飞行员数量最多的发展中国家之一。

III Write a summary of the text in 120 words.

Directions: A summary should be written in your own words. It should contain only the ideas of the original text. Do not insert any of your own opinions, interpretations, deductions or comments into the summary.

IV Read the passage, select one word for each blank from a list of choices in the bank, and change the form when necessary. You may not use any of the words in the bank more than once.

> lay off   gender   pandemic   at risk   elimination
> profession   initiative   recruit   maternity   impede

The path to the flight deck has never been easy for women. Beyond the 1) _____ assumptions, there are the structural forces 2) _____ progress. Male-dominated militaries have long fed pilots into airline cockpits, though vets have taken a smaller share of the openings in recent years. Once women do make it in, everything from a male-centered training environment to work rules concerning 3) _____ can slow their advancement.

There was reason for optimism before the 4) _____ hit. The share of female pilots rose to 5.3% worldwide this year, from 3% in 2016, according to the International Society of Women Airline Pilots in Las Vegas. Europe has led the way among the larger markets, along with a handful of countries like India, where the boom in air travel has attracted women to the well-paying 5) _____.

But with the novel coronavirus pandemic grounding lots of travel, that hard-won progress is 6) _____. Airlines are reeling, with more than 15,000 pilot jobs in

Europe alone under threat of 7) _____, according to the European Cockpit Association. American Airlines Group Inc. and other US carriers have started to 8) _____ tens of thousands of employees.

There's no active 9) _____ to attract more women because of the personnel surplus. EasyJet, which reached its target of 20% female new-entrant pilots this year, says it's paused to 10) _____ and it's not clear when it will begin taking on new staff again.

V Critical thinking

Work in groups and discuss the following questions.

Undoubtedly, gender equality has achieved quite a lot in the modern world. However, the gap between genders still exists. From the physical and psychological perspectives, do you think men pilots can do better than women pilots in aviation? Why or why not? Have you ever learned more about the development of gender equality? Do you think women's role in aviation is in accordance with it?

◉ **Further reading**

Munns' flying ambitions

1 According to the Aircraft Owners and Pilots Association (AOPA), female pilots "still account for only a small portion of the pilot population despite decades of attempts to fix the problem." The statistics back this up with female pilots currently "making up about just 7% of all certificated pilots and little has changed between 2010 and 2020 when the number was closer to 6.2%. Professional pilot ranks are even less diverse, with women holding 4.3% of airline transport pilot certificates." There has been some progress as noted by Peggy Chabrian who is the president and founder of Women in Aviation International (WAI), one of a few groups that advocate for a greater representation of women throughout the industry, especially in professional settings. Marnie Munns is one of the few female flight captains in the world. "Until a few years ago you could fit all the female captains into a 450-seater, A380," she says. But change is afoot.

2 Over the past three years she has been helping to cultivate a female empowerment program called the Amy Johnson Initiative with EasyJet, during which time the airline has increased recruitment of female cadet pilots from 5 to 13 percent. The aim is to reach 20 percent by 2020, "and we won't stop there," insists Munns.

3 With a long family history of air travel – and a daughter, aged six, who is currently toying with the career – Munns seems genetically predisposed to being airborne. However, women are still suspected to comprise only three percent of the pilot community globally, suggesting that while aviation technology has progressed, society's definition of a flight captain hasn't.

4 "I did have one gentleman who asked me if he could speak to the pilot," she recalls. And another time, "I had one lady who said she wasn't sure it was right to have two women pilots."

5 Such rare encounters are unfamiliar to Munns whose family never quashed or questioned her ambitions of flying. "Quite the opposite," says Munns. Her grandfather was a pilot in the RAF and her father, subsequently, hopped about Europe for work, often bringing young Marnie Munns with him on business trips. Looking back, Munns says it was no coincidence that she chose a career in aviation. "As a child, I would climb trees and just sit up there and read a book and listen to the leaves rustle and feel the breeze in my face."

6 Recently she also found a female wing of the Munns flying troupe. "It turns out my grandad's sister was also a pilot which was unusual at the time," she says. "She would save up her dress-making money to fund her flying lessons. Famous aviators like Amy Johnson were doing amazing things then," she says, referencing the first female pilot to fly from Britain to Australia – a personal icon for Munns – "but it was still quite rare to see women in the cockpit."

7 When Munns told her school career adviser of her sky-high ambitions, she was asked, "And what else?" To make sure no young person is dissuaded from following her flight path, Munns and her fellow female EasyJet pilots have traveled to more than 100 schools in the past year trying to embolden young women from all backgrounds to consider a career in aviation with EasyJet – a lucrative alternative to university fees.

8 "To become a pilot you might spend £110,000 on flying lessons, but you'd be guaranteed a job after 18 months and you'd start your job with £40,000 a year and going upwards," she says.

9 Why then are so few women choosing to work in aviation? Munns believes the problem lies in primary education. "It's been quite an eye opener, seeing my children go through the education system," she says. Surveys have shown that only 56 percent of girls enjoy taking part in sports lessons compared to 71 percent of boys. "Sport education is especially important in establishing an equal playing field," according to Munns. "Actually, that's where it all begins."

10 But even before they send their kids off to school, parents should keep their wits about them to avoid imparting subconscious prejudices. People are surprised, says Munns, by the flexible nature of the job which makes it an appealing prospect for those eager to build a family as well as a career. "My husband and I are both part-time pilots. I work the days when he's not working so we can manage the childcare between us. I think it's a very good choice for me and for my family. When I have days off, I can spend much more time with my kids," Munns said.

11 "If more people knew about flexible hours, perhaps parents wouldn't be so worried about their daughters going into the career and telling them 'you can't have a family,' because actually, aren't we all striving for that work-life balance? Do parents need to worry about their daughters if they choose to be a pilot?"

12 With the world at their fingertips, one might expect the Munns family to have jet-set holiday plans in abundance. But when they aren't floating in Earth's stratosphere, Munns and her husband like to take their two daughters on staycations: camping in the Cairngorms National Park for the last and campervanning planned for the next. "It's nice to do something a bit different," she says humbly.

13 Ultimately, Munns and her husband aren't pushing their daughters to work in aviation but, beyond the quantifiable successes of the Amy Johnson Initiative, working on the project brings its own personal reward as a parent. "I just want my children to be happy and the best that they can be and to not have anything stand in their way," she explains, "that's the most you can ask for."

14 It seems, just like her grandfather, Munns is poised to change the world, only this time, for a new generation of women who are ready to soar in the world of aviation.

New words

afoot /ə'fʊt/ *adj.* being planned or happening 计划中的; 进行中的

empowerment /ɪm'paʊəmənt/ *n.* the process of gaining freedom and power to do what you want or to control what happens to you 赋权, 使有能力

cadet /kə'det/ *n.* a young person who is training to become an officer in the police or armed forces（警校或军校的）学员

quash /kwɒʃ/ *v.* to take action to stop sth. from continuing 制止; 阻止

troupe /truːp/ *n.* a group of actors, singers, etc. who work together 剧团, 表演团

cockpit /'kɒk‚pɪt/ *n.* the area in a plane, boat or racing car where the pilot or driver sits（飞机的）驾驶舱;（船的）舵手座;（赛车的）驾驶座

dissuade /dɪ'sweɪd/ *v.* to persuade sb. not to do sth. 劝阻

embolden /ɪm'bəʊldən/ *v.* to make sb. feel braver or more confident 使勇敢,

使有胆量

lucrative /'luːkrətɪv/ *adj.* producing a large amount of money; making a large profit 赚大钱的; 获利多的

impart /ɪm'pɑːt/ *v.* to pass information, knowledge, etc. to other people 传授, 透露

subconscious /sʌb'kɒnʃəs/ *adj.* of or concerning the part of your mind that contains feelings that you are not aware of 下意识的, 潜意识的

stratosphere /'strætəsfɪə/ *n.* the layer of the Earth's atmosphere between about 10 and 50 kilometers above the surface of the Earth 平流层

staycation /steɪ'keɪʃən/ *n.* a holiday that you take at home or near your home rather than traveling to another place 在家中（或附近）度假（而不是到别处旅行）; 居家度假; 宅度假

quantifiable /'kwɒntɪ‚faɪəbəl/ *adj.* capable of being measured or described as a quantity 可以计量的

Phrases and expressions

back up 支持, 证实
keep one's wits 保持头脑清醒或冷静
strive for 力争

at one's fingertips 触手可及; 随时可供使用

Proper nouns

Aircraft Owners and Pilots Association
  (AOPA) 航空器拥有者及驾驶员协会
Women in Aviation International (WAI)
  国际航空妇女组织
Amy Johnson Initiative 埃米·约翰逊倡议

EasyJet 英国易捷航空公司
RAF (Royal Air Force) 英国皇家空军
Cairngorms National Park 凯恩戈姆
  斯国家公园（英国最大的国家公园）

Technical terms

A380 空客A380客机
cadet pilot 飞行学员

jet-set holiday plan 乘飞机旅行的假日
计划

Exercises

For each of the following unfinished statements or questions, choose the most appropriate answer from A, B, C, or D according to the text.

1   Why is Marnie Munns so special?

  A. She is a helicopter and fixed-wing pilot.

  B. She is the first flight instructor.

  C. She is one of the few female flight captains worldwide.

  D. She is the first female flight captain.

2   According to Marnie Munns, to _____ is her mission right now.

  A. increase the number of women in aviation

  B. teach young pilots how to operate aircraft

  C. introduce high school graduates to EasyJet

  D. diversify the face of women in the aviation industry

3   What's the family's influence on Marnie Munns?

  A. She wanted to become a doctor.

  B. She chose her own path as a female pilot.

  C. She was not interested in learning to fly.

  D. All of the above.

4   Few women chose to work in aviation because _____.

   A. girls don't like planes

   B. it's hard to keep work-life balance

   C. females are not good at flying

   D. there are problems in primary education

5   Is a pilot a good choice for those women who are eager to build a family as well as a career?

   A. No, they can't take care of the family.

   B. Yes, because being a pilot has a flexible nature.

   C. Yes, because they can earn a lot.

   D. No, they will get very busy.

Questions for discussion

1   What is the aim of the Amy Johnson Initiative?

2   Why does Munns seem genetically predisposed to being airborne?

3   Did Munns' school career adviser highly appreciate her sky-high ambitions? Why or why not?

4   What are the advantages of being a female pilot?

5   How does the Munns family spend their holidays?

⊙ **Practical writing**

How to illustrate data by using figures and tables

Figures play an amazingly important role in scientific publications, and no scientific paper could be written without figures. Because most scientific publications are supported by tremendous amounts of data, figures are always necessary. As a form of communication, figures (and in particular, the graphical display of quantitative data) can be used to convey information simply.

While statistical analysis sometimes reduces some data on the surface (expressing a mass of data by a few simple metrics), graphing contains the whole information of the data. Graphs can take advantage of the human brain's great power to recognize visual/spatial patterns and to quickly shift focus from pictures to small details.

But like all forms of communication, graphics can also confuse or deceive, to some extent. Thus, the rule of graphics is a simple one: They must help to reveal the truth. Just as disorganized writing often indicates disorganized thinking, a chart that fails to tell the story of the data usually means the author does not recognize what story should be told. Sufficient care should be given to the design and execution of graphics, just as in the design of the written paper itself.

How to develop effective tables and figures? Firstly, decide which results to present, paying attention to whether data are best presented either within the text as tables or figures. Then limit the number of tables and figures to those that provide essential information that could not adequately be presented in the text. Also, design each table and figure to be understandable on its own, without reference to the text. Next, number each figure and table in the order in which they are referred to in the text (figures and tables are numbered separately). Organize the tables and figures in such an order that they tell a story, but typically tables and figures are located on separate pages that follow the Reference section (depending on the targeted journal). Lastly, make sure there is no page break in the middle of a table or figure, if the journal wants the tables and figures integrated into the text. Do not wrap text around tables and figures and be sure all figures and tables are referenced in the text of the article, and also write the table titles and figure legends in the past tense.

Exercise

Design a table about how college students spend their pocket money or daily allowance according to the following paragraph and appreciate the amazing change brought by figures and a table.

College students have different preferences with respect to their pocket money or daily allowance. Freshmen mostly focus their concentration on food, accounting for 65%, and clothing, recreation and other items make up 8%, 3% and 24%, respectively. Sophomores spend 60% of their daily allowance on food and 9%, 5% and 26% on clothing, recreation and other items respectively. Juniors just spend 42% of their daily allowance on food, while they spend 15%, 8% and 35% on their clothing, recreation and other items respectively. Seniors spend least on their food, which only takes up 35% of their daily allowance, and they spend 17%, 10% and 38% on clothing, recreation and other items respectively.

Unit 5
Health risks of space travel

Learning objectives

Upon completion of this unit, you will be able to:

- recognize health problems that space exploration brings about;
- appraise astronauts' efforts to deal with the challenges and dangers;
- identify the elements that contribute to astronauts' survival;
- write an experiment report.

◉ Lead-in

There is no royal road to space exploration. As astronauts are faced with all kinds of challenges and dangers, they have to stand ready physically and mentally so as to ensure successful space missions. What are the possible challenges and dangers they might meet with? Do you think astronauts can serve as your role models when you come across setbacks? Why or why not?

⊙ Intensive reading

The challenges astronauts will face on a journey to Mars

1   NASA is trying to bring the various risks down before launching astronauts to Mars in the 2030s. The road to Mars is paved with peril. Astronauts on Red Planet missions will have to deal with deep-space radiation, the effects of microgravity and the stress of confinement and isolation, all at the same time and for a long, continuous period. It currently takes a minimum of six months to get to Mars after all, and just as long to get back. Crewmembers will have to make it through this gauntlet in good nick, both physically and mentally.

2   The spacecraft these astronauts launch aboard "will have to provide them with everything they need for basic survival, but even more than that, because we expect them to be capable of doing a job – a job that has mental demands, a job that has physical demands," said Jennifer Fogarty, the chief scientist with NASA's Human Research Program (HRP), during a presentation with the agency's Future In-Space Operations working group.

Many stressors

3   The HRP is tasked with characterizing the effects of spaceflight on astronauts and developing mitigation strategies. The program recognizes five classes of "stressor" that can significantly affect human health and performance on deep-space missions, Fogarty said. These are altered gravity fields, hostile closed environments, radiation, isolation/confinement, and distance from Earth (which means that help is very far away). HRP scientists and other researchers around the world are trying to get a handle on all of these stressors, by performing experiments here on Earth and carefully monitoring the mental and physical health of astronauts living on the International Space Station (ISS).

4   The long-term goal of such work is to help enable crewed missions to Mars, which NASA wants to pull off before the end of the 2030s. Indeed, a few years ago, NASA astronaut Scott Kelly and cosmonaut Mikhail Korniyenko stayed aboard the ISS for 11 months – about twice as long as the usual time – to help researchers measure the impact of very long space missions, such as the roundtrip journey to Mars.

5   "It's tough to accurately characterize the toll that such a voyage will take on an astronaut, however. That's because the cumulative effect of the spaceflight stressors might be additive or interactive," Fogarty said, "and putting all of the hazards together in an experimental setting is nearly impossible." For example, scientists perform radiation studies on lab animals here on Earth. But microgravity isn't part of that experimental picture, and adding it to the mix isn't feasible at the moment. The ISS cannot provide deep-space radiation data, because it orbits within Earth's protective magnetosphere. And installing radiation-emitting equipment aboard the orbiting lab doesn't seem like a great idea.

Biggest concerns

6   Some of the stressors are more worrying than others. For example, researchers and NASA officials have repeatedly cited radiation as one of the biggest Mars-mission hazards. High radiation exposure increases astronauts' risk of developing cancer later in life, but there are more immediate concerns as well. For instance, a recent study determined that crewmembers on a Red Planet mission will likely receive cumulative doses high enough to damage their central nervous systems. Astronauts' moods, memory and learning ability may be harmed as a result, the study found.

7   Fogarty mentioned another issue that requires focused research attention – spaceflight associated neuro-ocular syndrome (SANS), also known as visual impairment and intracranial pressure (VIIP) syndrome. It describes the potentially significant and long-lasting vision problems that spaceflight can cause in astronauts, likely because of fluid shifts that increase pressure inside the skull. SANS "right now in low Earth orbit is very, very manageable and recoverable, but we don't know

the system well enough to predict whether it will remain that way for something like an exploration mission," Fogarty said. "So, this is one of our highest-priority physiological areas that we're studying right now."

The moon as a proving ground

8 NASA isn't planning to go straight to Mars. The agency aims to land two astronauts near the lunar south pole by 2024, then establish a long-term, sustainable presence on and around the moon shortly after that time.

9 Indeed, the main goal of these activities, which NASA will conduct via a program called Artemis, is to learn the skills and techniques needed to send astronauts to Mars, agency officials have said. One of Artemis' key pieces of infrastructure is a small moon-orbiting space station called the Gateway, which will serve as a hub for surface activities. For example, landers, both robotic and crewed, will descend toward the lunar surface from Gateway, and astronauts aboard the outpost will likely operate rovers from up there as well, NASA officials have said.

10 A great deal of research will be conducted on Gateway as well, and much of it will investigate astronauts' health and performance in a true deep-space environment. Fogarty mentioned one research strategy that may be particularly useful to planners mapping out the path to Mars – studying small samples of human tissue aboard the moon-orbiting outpost. Such work will help researchers get around one of the biggest issues affecting studies that use rodents and other non-human animals as model organisms, Fogarty said – that of "translatability". "How do we bridge the difference between a rat or a mouse and a human? Because it's not directly applicable, and that's plaguing terrestrial medicine and research as well," she said.

11 "But with the invention of, and the continued validation of, organs and tissue on a chip – those are actual human organs and tissue, you can connect them, and essentially you can recapitulate very sophisticated aspects of a human using these chips," Fogarty added, "I think we can make significant progress understanding the complex environment using the chip as a model organism to interpret really where we're going with the human limitation."

New words

crewmember /ˈkruːˌmembə/ *n.* a member of a group of people who work together esp. on a ship 工作人员；（尤指）船员

stressor /ˈstresə/ *n.* an event, experience, etc. that causes stress 导致压力的因素

mitigation /ˌmɪtɪˈɡeɪʃən/ *n.* a reduction in the unpleasantness, seriousness, or painfulness of sth. 减轻，缓和

cumulative /ˈkjuːmjɡlətɪv/ *adj.* increasing gradually as more of sth. is added or happens 积累的，渐增的

magnetosphere /mæɡˈniːtəsfɪə/ *n.* the region surrounding a planet, such as the Earth, in which the behavior of charged particles is controlled by the planet's magnetic field 磁气圈，磁层

recoverable /rɪˈkʌvərəbəl/ *adj.* (of sth. lost) able to be regained or retrieved 可恢复的；可重新获得的

sustainable /səˈsteɪnəbəl/ *adj.* able to continue or be continued for a long time 可持续的

lander /ˈlændə/ *n.* a spacecraft designed to land on the surface of a planet or moon 着陆器

rodent /ˈrəʊdənt/ *n.* a small mammal which has sharp front teeth 啮齿动物

plague /pleɪɡ/ *v.* to cause worry, pain, or difficulty to someone or something over a period of time 不断困扰

terrestrial /tɪˈrestriəl/ *adj.* relating to the Earth rather than to the moon or other planets 地球的；与地球有关的

validation /ˌvælɪˈdeɪʃən/ *n.* the act of proving that sth. is true 证实；认可

recapitulate /ˌriːkəˈpɪtʃɡleɪt/ *v.* to repeat or give a summary of what has already been said, decided, etc. 扼要重述；概括

Phrases and expressions

in good nick 状态良好
be tasked with doing sth. 被指派去做某事
get a handle on 开始掌握
pull off 成功完成

take a toll on 对⋯造成严重损害
conduct research on 对⋯进行研究
map out 筹划
get around 克服

Proper nouns

Red Planet 火星，红色星球，红色行星	Future In-Space Operations 未来太空行动

Technical terms

deep-space radiation 深空辐射	experimental setting 实验环境
gravity field 重力场	protective magnetosphere 保护磁层
roundtrip journey 往返行程	central nervous system 中枢神经系统
cumulative effect 累积效应	proving ground 试验场所

Background information

Human Research Program (HRP) 人类研究项目

The Human Research Program (HRP) investigates and mitigates the highest risks to human health and performance, providing essential countermeasures and technologies for human space exploration. Risks include physiological and performance effects from hazards such as radiation, altered gravity, and hostile environments, as well as unique challenges in medical support, human factors, and behavioral health support.

spaceflight associated neuro-ocular syndrome (SANS) 太空飞行相关神经视觉综合征

Spaceflight associated neuro-ocular syndrome (SANS), formerly called visual impairment and intracranial pressure (VIIP) syndrome (视力障碍和颅内压综合征), is a constellation of findings and symptoms that have been found in astronauts who have undergone long-duration spaceflight missions in microgravity environments.

Artemis "阿尔忒弥斯" 计划

With the Artemis program, NASA will land the first woman and first person of color on the moon by 2024, using innovative technologies to explore more of the lunar surface than ever before. It will collaborate with its commercial and international partners and establish sustainable exploration on the moon. Then, it will use what is learned on and around the moon to take the next giant leap – sending astronauts to Mars.

Gateway "门户" 月球轨道空间站

The Gateway, a vital component of NASA's Artemis program, will serve as a multi-purpose outpost orbiting the moon that provides vital support for a sustainable, long-term human return to the lunar surface and serves as a staging point for deep space exploration. NASA is working with commercial and international partners to establish the Gateway.

D ifficult sentences

1 **NASA is trying to bring the various risks down before launching astronauts to Mars in the 2030s. (Para. 1)**

NASA is going to send astronauts to Mars in the 2030s. But before that, it is trying to reduce all kinds of risks that astronauts might face in the future.

2 **Crewmembers will have to make it through this gauntlet in good nick, both physically and mentally. (Para. 1)**

Astronauts will face very direct and hazardous challenges on their journey, as a soldier would face them in a battle. But the hope is that they will finish the task in good physical and mental condition.

3 **The HRP is tasked with characterizing the effects of spaceflight on astronauts and developing mitigation strategies. (Para. 3)**

The HRP has been given a task to study the dangerous influences of spaceflight on astronauts and find solutions to relieve the problems.

4 **The long-term goal of such work is to help enable crewed missions to Mars, which NASA wants to pull off before the end of the 2030s. (Para. 4)**

The ongoing purpose of the work is to finally make manned spaceflight to Mars possible, and NASA wants to succeed in doing so before the end of the 2030s.

5 **Fogarty mentioned one research strategy that may be particularly useful to planners mapping out the path to Mars – studying small samples of human tissue aboard the moon-orbiting outpost. (Para. 10)**

Fogarty said it is a good idea to study small samples from the human body when they are in the moon-orbiting outpost. It may be very helpful to the plan of sending humans to Mars in future.

6 **Such work will help researchers get around one of the biggest issues affecting studies that use rodents and other non-human animals as model organisms, Fogarty said – that of "translatability". (Para. 10)**

One strategy is to use rodents, such as rats and mice, to do experiments in an environment beyond Earth. However, rodents are not humans. And the results have to be transferred to apply to human organisms, though doing experiments in a moon-orbiting outpost will help.

7　**"How do we bridge the difference between a rat or a mouse and a human? Because it's not directly applicable, and that's plaguing terrestrial medicine and research as well," she said. (Para. 10)**

"While we conduct medical experiments on rodents, such as mice or rats, the results still need to be applied to humans. That's a problem, because humans are not rodents. This is the same challenge facing medical research and medicine here on Earth," she said.

8　**"I think we can make significant progress understanding the complex environment using the chip as a model organism to interpret really where we're going with the human limitation." (Para. 11)**

"In my opinion we can move forward in understanding the complicated environment. With the help of organs and tissue on a chip, the researchers can carry out experiments and understand where the limits of humans are."

Exercises

I　**Fill in the blanks with the words and expressions given below, and change the forms when necessary.**

cumulative　survival　infrastructure　get around　mitigation
sophisticated　recoverable　sustainable　interpret　mission

1　The Long March 9 will not only be used for lunar _____ but also be required for other deep-space exploration projects.

2　I think we should be able to _____ most of these problems with the help of our professor.

3　If the program does not indicate this, a user will _____ it as rudeness at best; at worst, he will assume the program has crashed and that drastic action must be taken.

4   The administration will coordinate with different government departments to include the safety of heritage sites in the national disaster prevention and _____ system.

5   The main reason many amateur photographers upgrade from their handy compact digital cameras to _____ digital SLR versions is because they are tired of dealing with the dreaded lag time.

6   Lack of water and oxygen poses a serious threat to the _____ of astronauts.

7   _____ investment can create jobs and lay the foundation for future economic growth.

8   Removal expenses might be _____ if you have to move to a different area in order to find work.

9   At the summit, a plan was put forward that the government would import commodities with a(n) _____ value of over $170 billion from European countries in the coming five years.

10 According to the scientist, many of the crops grown for ethanol are not done so in a(n) _____ manner, resulting in habitat destruction and loss of valuable food supply resources.

‖ Translation

Part A   Translate the following paragraph into Chinese.

The HRP is tasked with characterizing the effects of spaceflight on astronauts and developing mitigation strategies. The program recognizes five classes of "stressor" that can significantly affect human health and performance on deep-space missions, Fogarty said. These are altered gravity fields, hostile closed environments, radiation, isolation/confinement, and distance from Earth (which means that help is very far away). HRP scientists and other researchers around the world are trying to get a handle on all of these stressors, by performing experiments here on Earth and carefully monitoring the mental and physical health of astronauts living on the International Space Station (ISS).

Part B   Translate the following paragraph into English.

中国国内研制的宇航服（spacesuit）确保了宇航员在空间站天和核心舱逗留期间以及在空间站外活动时的安全。他们有舱内（intravehicular）宇航服和舱外（extravehicular）宇航服。舱内宇航服保证了正常情况下的通风（ventilation）和散热（heat dissipation），并在航天器发生泄漏时提供氧气以确保宇航员的安全。舱外宇航服为在航天器外工作的宇航员提供安全有效的环境保护、环境控制和生命保障。

||| Write a summary of the text in 120 words.

Directions: A summary should be written in your own words. It should contain only the ideas of the original text. Do not insert any of your own opinions, interpretations, deductions or comments into the summary.

|V Read the passage, select one word for each blank from a list of choices in the bank, and change the form when necessary. You may not use any of the words in the bank more than once.

complete	harmful	prolong	physical	further
spaceflight	impact	astronaut	agency	microgravity

The effects of 1) _____ on the human body are well known, with 2) _____ seeing some changes to bone density, muscle strength and other 3) _____ changes, such as puffy (肿胀的) faces or changes to their eyes. But a new study suggests the changes may go even 4) _____ – they may cause harm to the immune system and lead to cancer.

The study, which was published in November 2018 in the *Journal of Applied Physiology*, highlights the negative 5) _____ on natural killer cells for astronauts who have spent approximately six months on the International Space Station.

"What NASA and other space 6) _____ are concerned about is whether or not the immune system is going to be weakened during very 7) _____ spaceflight missions," said Richard Simpson, senior author and associate professor of nutritional sciences at the University of Arizona, in a statement. "What clinical risks are there to the astronauts during these missions when they're exposed to things like 8) _____, radiation and isolation stress? Could it be 9) _____ to the level that the astronaut wouldn't be able to 10) _____ the mission?"

V Critical thinking

Work in groups and discuss the following questions.

It is said that there is no royal road to space exploration. Astronauts have to deal with the adverse effects of radiation, microgravity, isolation, boredom on a long mission, etc. Imagine you, as an astronaut, would spend six months on Tiangong space station. One of the problems that you would face on the long mission is boredom. How would you combat boredom in space? What activities would help to keep you fit both physically and mentally?

◉ **Further reading**

Five hazards of human spaceflight

1 A human journey to Mars, at first glance, offers an inexhaustible amount of complexities. To bring a mission to the Red Planet from fiction to fact, NASA's Human Research Program has organized hazards astronauts will encounter on a continual basis into five classifications. Pooling the challenges into categories allows for an organized effort to overcome the obstacles that lay before such a mission. However, these hazards do not stand alone. They can feed off one another and exacerbate effects on the human body. These hazards are being studied using ground-based analogs, laboratories, and the International Space Station, which serves as a test bed to evaluate human performance and countermeasures required for the exploration of space.

2 Various research platforms give NASA a valuable insight into how the human body and mind might respond during extended forays into space. The resulting data, technology and methods developed serve as valuable knowledge to extrapolate to multi-year interplanetary missions.

Radiation

3 The first hazard of a human mission to Mars, radiation, is also the most difficult to visualize because, space radiation is invisible to the human eye. Radiation is not only stealthy, but considered one of the most menacing of the five hazards. Outside Earth's natural protection, radiation exposure increases cancer risk, damages the central nervous system, and can alter cognitive function, reduce motor function and prompt behavioral changes. To learn what can happen above low Earth orbit, NASA studies how radiation affects biological samples using a ground-based research laboratory. The space station sits just within Earth's protective magnetic field, so while our astronauts are exposed to ten-times higher radiation than on Earth, it's still a smaller dose than what deep space has in store. To mitigate this hazard, deep space vehicles

will have significant protective shielding, dosimetry, and alerts. Research is also being conducted in the field of medical countermeasures such as pharmaceuticals to help defend against radiation.

Isolation

4 Behavioral issues among groups of people crammed in a small space over a long period of time, no matter how well trained they are, are inevitable. Crews will be carefully chosen, trained and supported to ensure they can work effectively as a team for months or years in space.

5 On Earth we have the luxury of picking up our cell phones and instantly being connected with nearly everything and everyone around us. On a trip to Mars, astronauts will be more isolated and confined than we can imagine. Sleep loss, circadian desynchronization, and work overload compound this issue and may lead to performance decrements, adverse health outcomes, and compromised mission objectives.

6 To address this hazard, methods for monitoring behavioral health and adapting/refining various tools and technologies for use in the spaceflight environment are being developed to detect and treat early risk factors. Research is also being conducted in workload and performance, light therapy for circadian alignment, phase shifting and alertness.

Distance

7 The third and perhaps most apparent hazard is, quite simply, the distance. Mars is, on average, 225 million kilometers from Earth. Rather than a three-day lunar trip, astronauts would be leaving our planet for roughly three years. While International Space Station expeditions serve as a rough foundation for the expected impact on planning logistics for such a trip, the data isn't always comparable. If a medical event or emergency happens on the station, the crew can return home within hours. Additionally, cargo vehicles continually resupply the crews with fresh food, medical equipment, and other resources. Once you burn your engines for Mars, there is no turning back and no resupply.

8   Planning and self-sufficiency are essential keys to a successful Martian mission. Facing a communication delay of up to 20 minutes one way and the possibility of equipment failures or a medical emergency, astronauts must be capable of confronting an array of situations without support from their fellow team on the Earth.

Gravity

9   The variance of gravity that astronauts will encounter is the fourth hazard of a human mission. On Mars, astronauts would need to live and work in three-eighths of Earth's gravitational pull for up to two years. Additionally, on the six-month trek between the planets, explorers will experience total weightlessness.

10 Besides Mars and deep space there is a third gravity field that must be considered. When astronauts finally return home they will need to readapt many of the systems in their bodies to Earth's gravity. Bones, muscles, cardiovascular system have all been impacted by years without standard gravity. To further complicate the problem, when astronauts transition from one gravity field to another, it's usually quite an intense experience. Blasting off from the surface of a planet or a hurdling descent through an atmosphere is many times the force of gravity.

11 Research is being conducted to ensure that astronauts stay healthy before, during and after their mission. NASA is identifying how current and future, FDA-approved osteoporosis treatments, and the optimal timing for such therapies could be employed to mitigate the risk for astronauts developing premature osteoporosis. Adaptability training programs and improving the ability to detect relevant sensory input are being investigated to mitigate balance control issues. Research is ongoing to characterize optimal exercise prescriptions for individual astronauts, as well as defining metabolic costs of critical mission tasks they would expect to encounter on a Mars mission.

Closed environments

12 A spacecraft is not only a home; it's also a machine. NASA understands that the ecosystem inside a vehicle plays a big role in everyday astronaut life. Important

habitability factors include temperature, pressure, lighting, noise, and quantity of space. It's essential that astronauts are getting the requisite food, sleep and exercise needed to stay healthy and happy.

13 Technology, as often is the case with out-of-this-world exploration, comes to the rescue in creating a habitable home in a harsh environment. Everything is monitored, from air quality to possible microbial inhabitants. Microorganisms that naturally live on your body are transferred more easily from one person to another in a closed environment. Astronauts, too, contribute data points via urine and blood samples, and can reveal valuable information about possible stressors. The occupants are also asked to provide feedback about their living environment, including physical impressions and sensations so that the evolution of spacecraft can continue addressing the needs of humans in space. Extensive recycling of resources we take for granted is also imperative: oxygen, water, carbon dioxide, even our waste.

Human research essential to space exploration

14 NASA has already gone beyond simply identifying five challenges of human spaceflight to facilitate a focused and organized effort to reach Mars. Within the agency, there are entities dedicated to the evolution of spaceflight in all five of these areas.

15 NASA's Human Research Program (HRP) remains committed to preserving the health and vitality of the crew that will someday touch down upon Mars. It is dedicated to discovering the best methods and technologies to support safe, productive human space travel. HRP enables space exploration by reducing the risks to astronaut health and performance using ground research facilities, the International Space Station, and analog environments. This leads to the development and delivery of an exploration biomedical program focused on: informing human health, performance, and habitability standards; the development of countermeasures and risk mitigation solutions; and advanced habitability and medical support technologies.

16 While these five hazards present significant challenges, they also offer opportunities for growth and innovation in technology, medicine and our understanding of the human body. One human challenge explored, one step closer to Mars.

New words

exacerbate /ɪɡˈzæsəbeɪt/ *v.* to make sth. worse, especially a disease or problem 使恶化；使加重

analog /ˈænəlɒɡ/ *n.* a thing that is similar to another thing 相似物，类似物

extrapolate /ɪkˈstræpəleɪt/ *v.* to estimate sth. or form an opinion about sth., using the facts that you have now and that are relevant to one situation and supposing that they will be relevant to the new one 推断，推知

dosimetry /dəʊˈsɪmətri/ *n.* a method of measuring the dose of radiation emitted by a radioactive source 放射量测定；剂量测定

circadian /sɜːˈkeɪdiən/ *adj.* connected with the changes in the bodies of people or animals over each period of 24 hours 昼夜节律的，生理节奏的

decrement /ˈdekrəmənt/ *n.* a reduction, or a lower level or amount 消耗；减缩

alignment /əˈlaɪnmənt/ *n.* an arrangement in which two or more things are positioned in a straight line or parallel to each other 排成直线；平行的行列

logistics /ləˈdʒɪstɪks/ *n.* the practical organization that is needed to make a complicated plan successful when a lot of people and equipment are involved 统筹安排

cardiovascular /ˌkɑːdiəʊˈvæskjələ/ *adj.* relating to the heart and the blood vessels (= the tubes that carry blood around the body) 心血管的

osteoporosis /ˌɒstiəʊpəˈrəʊsɪs/ *n.* a condition in which the bones become weak and are easily broken, usually when people get older or because they do not eat enough of certain substances 骨质疏松（症）

metabolic /ˌmetəˈbɒlɪk◄/ *adj.* connected with the chemical processes in living things that change food, etc. into energy and materials for growth 新陈代谢的

Phrases and expressions

stand alone 孤立存在

give sb. an insight into 给某人提供关于…的深刻见解

be exposed to 暴露在…中

adverse health outcomes 不良健康后果

further complicate the problem 使问题更复杂

optimal timing 最佳时机

Proper nouns

FDA (Food and Drug Administration) 美国食品和药物管理局

Technical terms

protective magnetic field 保护性磁场   同步化, 生理节奏紊乱
circadian desynchronization 昼夜节律去   variance of gravity 重力变化

Exercises

For each of the following unfinished statements or questions, choose the most appropriate answer from A, B, C, or D according to the text.

1   Radiation ranks first among the five hazards of spaceflight because of its
    _____.

    A. availability

    B. visibility

    C. invisibility

    D. odd smell

2   Which plight is NOT mentioned for astronauts on a trip to Mars?

    A. Performance reductions.

    B. Discounted mission goals.

    C. Unfavorable health results.

    D. Hearing problems.

3   Astronauts should worry about important habitability factors EXCEPT
    _____.

    A. temperature

    B. space

    C. exercise

    D. recreation

4   How can the astronauts contribute to satisfying the needs of humans in space?

   A. By providing data about their own physical status.

   B. By keeping communication with Earth.

   C. By neglecting the daily recycling of life.

   D. By lowering the need of the requisite food.

5   Which of the following is true according to the text?

   A. The professional training of crews will mainly focus on health and
      performance.

   B. Muscular problems should be given much thought due to their high
      incidence for astronauts in space.

   C. Not everything in the closed space shuttle will be carefully monitored.

   D. The HRP mainly offers financial aids to relevant organizations in the US.

Questions for discussion

1   What will be the different effects of radiation on the astronauts in a
    ground-based research lab?

2   How long does it take for the crew to return home if emergencies arise
    on the International Space Station?

3   What gravity fields will astronauts traveling to Mars be concerned about?

4   What activity will greatly increase the intensity of gravity?

5   What will be more possibly transmitted in a closed setting?

◉ Practical writing

How to write an experiment report

An experiment report is an essential part of all laboratory courses and usually a significant part of your grade. It is intended to explain what you did in your experiment, what you learned, what the results meant, etc. The report should contain at least six parts: abstract, introduction, purpose/problem, procedure, observations and data, and conclusion.

Abstract

Remember to include an abstract in your experiment report. Write it as one paragraph. This is a concise summary of the entire experiment, which includes the rationale, methods, results, and significance in a brief form.

Introduction

Briefly state what has been known or found out about the problem so that you can be fully aware of what you should focus on in the experiment. This is background information from books, teachers, or other sources, giving readers an overall understanding of possible principles and information about the experiment.

Purpose/Problem

The purpose or problem deals with why you are conducting the experiment. You should write down the exact problem that will be investigated or dealt with.

Procedure

The procedure demonstrates accurately what was done. Make it more specific. As the procedure affects the results, it is essential to accurately explain what was done in the experiment.

Observations and Data

The observations present exactly what happened when the lab experiment was underway. An observation is measurable information that can be observed through your senses. Results include experimental data in the form of tables, graphs, drawings and other observations. Observations and data should be included in this part, as well as calculations made during the experiment.

Conclusion

Review and summarize briefly what was done in this experiment and what was found in the results. State in general terms the most important discoveries. You may also suggest follow-up experiments, and make some predictions according to what has been discovered.

Exercise

Write an English experiment report relevant to your major. The report should cover the six parts introduced above.

Mysteries of space

Learning objectives

Upon completion of this unit, you will be able to:

- get insights into some interesting physical phenomena in deep space;
- analyze relationships between objects from a scientific perspective;
- develop the spirit of exploration into the unknown;
- make a science and technology investigation report.

◉ Lead-in

Humanity's interest in the heavens has been universal and enduring. Humans are driven to explore the unknown, discover new worlds, push the boundaries of our scientific and technical limits, and then push further. When staring into starry nights, have you ever wondered what lies beyond those stars that twinkle like diamonds in deep space? Have you ever imagined what kind of experience people would have in a black hole? What can common people do to relate their daily life to space resources? Read on and find more, brave explorers.

◉ Intensive reading

What's it like inside a black hole?

1 You've managed to travel tens of thousands of light-years beyond the solar system. Bravely facing the depths of the great interstellar voids, you've witnessed some of the most achingly beautiful and outrageously powerful events in the universe, from the births of new solar systems to the cataclysmic deaths of massive stars. And now for your swan song, you're going big: You're about to take a dip into the inky blackness of a giant black hole and see what's on the other side of that enigmatic event horizon. What will you find inside? Read on, brave explorer.

Nearing the monster

2 First, we need to clear up some definitions. There are many kinds of black holes: some big ones, some small ones, some with electric charge, some without, and some with rapid rotations and others more sedentary. For the purposes of our adventure in this particular tale, I'm going to stick to the simplest possible scenario: a giant black hole with no electric charge and no spin whatsoever. Of course, this is decidedly unrealistic, but it's still a fun story with plenty of cool physics to unpack. We can save a more realistic trip for another visit, assuming we'll survive this hypothetical journey into a black hole, which of course we won't.

3 From a distance the black hole is surprisingly benign. After all, it's just a massive object, pretty much like any other massive object. Gravity is gravity and mass is mass – a black hole with the mass of, say, the sun will pull on you exactly the same as the sun itself. All that's missing is the wonderful heat and light and radiation. But if you felt like orbiting it at a safe distance, you most certainly could.

4 But why bother orbiting it when you could go farther in? The black hole itself is a singularity, a point of infinite density. But you can't see the singularity itself; it's shrouded by the event horizon, what we generally and wisely consider the "surface" of the black hole. To go farther, you must first pierce that veil.

Beyond the horizon

5   The event horizon isn't a real, physical boundary. It's not a membrane or a surface. It's simply defined as a particular distance from the singularity, the distance where if you fall below this threshold, you can't get out. You know, no big deal.

6   This is the distance from the singularity where the gravitational pull is so extreme that nothing, not even light itself, can escape the black hole's clutches. If you were to fall below this boundary and decided you had enough of this black hole exploration business, then too bad. As hard as you fired your rockets, you would find yourself no farther from the singularity. You're trapped, doomed.

7   But not instantly. You have a few moments to enjoy the experience before you meet your inevitable demise, if "enjoy" is the right word. How long it takes to reach the singularity depends on the mass of the black hole. For a small black hole (a few times the mass of the sun counts as "small") you can't even blink an eye. For a giant one, at least a million times bigger than our sun, you have a handful of heartbeats to experience this mysterious corner of the universe.

8   But hit the singularity you must. You don't get a choice. Within the event horizon, nothing can stay still. You are forever compelled to move. And the singularity lies in all your possible futures. Outside the black hole's event horizon, you can move in any direction in space you please. Up? Left? A little bit of both? Neither? The choice is yours. But no matter where you go or don't go in space, you must always travel into your future. You simply can't escape it. Inside the event horizon of a black hole, this common-sense understanding breaks down. Here, a single point – the singularity – lies in your future. You simply must travel toward the singularity. Turn left, turn up, turn around, it doesn't matter – the singularity always remains in front of you. And you will hit that singularity in a finite amount of time.

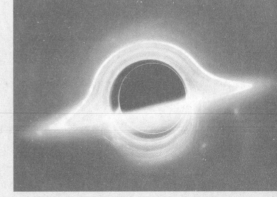

A rendezvous with infinity

9   As you fall toward the singularity, you're not cloaked in blackness. Light from the surrounding universe fell

in with you and continues to fall in after you. Due to the extreme gravity, that light is shifted to higher frequencies, and because of time dilation the outside universe appears sped up, but it's still there.

10 That's not to say it isn't weird. Because all the mass of the black hole is concentrated into an infinitely tiny point, the differences in gravity are extreme. You are stretched head to toe in an aptly named process known as spaghettification. And what's more, you're squeezed along your midsection. This squeezing operates on the beams of light surrounding you as well, concentrating the infalling light into a bright band about your waist.

11 Your view of the singularity becomes grotesque and distorted as well. It's pitch black – you can't see it, because it lies in your future, and just like your future you don't know what it looks like until you get there. But instead of appearing as a tiny point, the huge gravitational differences stretch that point to engulf most of your vision.

12 As you approach the singularity, it appears as if you're landing on the surface of a vast, featureless, empty, black planet. When the singularity stretches completely from horizon to horizon, then you've made it.

New words

interstellar /ˌɪntəˈstelə◂/ *adj.* between the stars in the sky 星际的

outrageously /aʊtˈreɪdʒəsli/ *adv.* in a very amazing manner 令人惊讶地

cataclysmic /ˌkætəˈklɪzəmɪk/ *adj.* causing a lot of destruction, or a sudden, violent change 灾难性的; 剧变的

inky /ˈɪŋki/ *adj.* black or very dark 漆黑的; 墨黑的

enigmatic /ˌenɪɡˈmætɪk◂/ *adj.* mysterious and difficult to understand 神秘的; 费解的

charge /tʃɑːdʒ/ *n.* the amount of electricity that an electrical device stores or that a substance carries 电荷

sedentary /ˈsedəntəri/ *adj.* involving a lot of sitting or little activity 静坐的; 久坐不动的

scenario /səˈnɑːriəʊ/ *n.* a situation that could possibly happen 可能发生的事; 可能出现的情况

decidedly /dɪˈsaɪdɪdli/ *adv.* definitely or in a way that is easily noticed 肯定地, 无疑地; 明显地

unpack /ʌnˈpæk/ *v.* to separate sth. into parts so that it is easier to understand 剖析, 分析

hypothetical /ˌhaɪpəˈθetɪkəl◂/ *adj.* based on possible ideas or situations rather than actual ones 假设的, 假定的

benign /bɪˈnaɪn/ *adj.* kind and gentle 和善的; 温和的

shroud /ʃraʊd/ *v.* to cover or hide sth. 笼罩; 覆盖

pierce /pɪəs/ *v.* to make a small hole in sth., or to go through sth., using a sharp object 刺破; 穿透

membrane /ˈmembreɪn/ *n.* a very thin layer of material that covers something 膜状物

threshold /ˈθreʃhəʊld/ *n.* the level or point at which you start to experience something, or at which something starts to happen or have an effect 阈, 界限; 起点

clutch /klʌtʃ/ *n.* a tight grasp or an act of grasping something 紧握; 掌控

demise /dɪˈmaɪz/ *n.* death 死亡, 逝世

cloak /kləʊk/ *v.* to cover sth. 笼罩, 遮盖

aptly /ˈæptli/ *adv.* in a way that is suitable or right for a particular situation 恰当地, 适宜地

grotesque /ɡrəʊˈtesk/ *adj.* strange or unusual in a way that is shocking or offensive 怪诞的; 怪异的

engulf /ɪnˈɡʌlf/ *v.* to surround or to cover sth. completely 完全包围, 吞没

Phrases and expressions

take a dip into 遨游; 涉猎	be compelled to 不得不; 被迫
clear up 解释; 澄清	break down 崩溃; 失败; 出故障
stick to 坚持; 信守, 遵守	concentrate … into 集中…于

Technical terms

solar systems 类太阳系

Background information

black hole 黑洞
A black hole is a region in space where the pulling force of gravity is so strong that light is not able to escape. The strong gravity occurs because matter has been pressed into a tiny space. This compression can take place at the end of a star's life. Some black holes are a result of dying stars.

swan song 天鹅之歌 (喻指艺术家的最后杰作)
A swan song is used to refer to the last piece of work produced by an artist, musician, etc., or the last performance by an actor, singer, etc.

event horizon 事件视界
Event horizon refers to the theoretical boundary surrounding a black hole, within which gravitational attraction is so great that nothing, not even radiation, can escape because the escape velocity is greater than the speed of flight.

singularity 奇点
A singularity refers to a hypothetical region in space in which gravitational forces cause matter to be infinitely compressed and space and time to become infinitely distorted.

time dilation 时间膨胀
In physics and relativity, time dilation is the difference in the elapsed time as measured by two clocks. It is either due to a relative velocity between them (special relativistic "kinetic" time dilation) or to a difference in gravitational potential between their locations (general relativistic gravitational time dilation).

spaghettification 面条化 (亦称 "意大利面化" 或 "意大利面效应")
This term is used to refer to the process by which (in some theories) an object would be stretched and ripped apart by gravitational forces on falling into a black hole.

D ifficult sentences

1   **Bravely facing the depths of the great interstellar voids, you've witnessed some of the most achingly beautiful and outrageously powerful events in the universe, from the births of new solar systems to the cataclysmic deaths of massive stars. (Para. 1)**

Exploring the vast emptiness between and beyond stars in deep space with great courage, you've experienced some of the most extremely beautiful and powerful events. From those events, you have come to know how some new solar systems have come into being and how some huge stars have been destroyed.

2   **And now for your swan song, you're going big: You're about to take a dip into the inky blackness of a giant black hole and see what's on the other side of that enigmatic event horizon. (Para. 1)**

And now you are going to do something fantastically great in the last part of your journey. That is to go into a huge black hole and find out what it is like in the completely unknown world.

3   **For the purposes of our adventure in this particular tale, I'm going to stick to the simplest possible scenario: a giant black hole with no electric charge and no spin whatsoever. (Para. 2)**

To make our trip to a black hole easy, I'd like to keep to the simplest possible setting. That is to say, the giant black hole we are going to visit has no electric charge and definitely stays still, not moving at all.

4   **Of course, this is decidedly unrealistic, but it's still a fun story with plenty of cool physics to unpack. We can save a more realistic trip for another visit, assuming we'll survive this hypothetical journey into a black hole, which of course we won't. (Para. 2)**

Of course, it is not realistic for us to find a black hole without electric charge or motion, but we will still find this kind of story to be fun and an interesting exploration of the physical field. Suppose we are still alive after this journey, we can take a more realistic trip, but of course we won't make it because we won't survive this trip.

5 It's simply defined as a particular distance from the singularity, the distance where if you fall below this threshold, you can't get out. (Para. 5)

The event horizon is simply used to refer to a particular distance from a point of infinite density which marks the boundary of a black hole. And then, if you go beyond to the other side of the boundary, you would be in the black hole and would never get out again.

6 Inside the event horizon of a black hole, this common-sense understanding breaks down. (Para. 8)

Once you come within the event horizon of a black hole, all your normal understanding about space travel wouldn't work any longer.

7 Due to the extreme gravity, that light is shifted to higher frequencies, and because of time dilation the outside universe appears sped up, but it's still there. (Para. 9)

Because the black hole's pulling force that attracts objects is extremely powerful, the rate at which the light vibrates becomes higher. At the same time, because of the effect of the slowing of time in the gravitational field, the outside universe appears to have been sped up. But, anyway, the universe still exists where it is.

Exercises

| Fill in the blanks with the words and expressions given below, and change the forms when necessary.

> scenario   enigmatic   outrageously   escape   shroud
> engulf   decidedly   sedentary   pierce   demise

1 The Black Death, also known as the Great Plague, spread across Europe between 1346 and 1353 and brought a swift but very painful _____ to up to 60 percent of the population – perhaps 200 million people in all.

2 There exists a tendency that houses in areas of rapidly economic development were _____ priced and that they are still priced much higher than the national average.

3 A snowstorm _____ the northern areas of the country last night.

4 If we persist in developing the economy regardless of the environment, we can predict the possible _____: the Earth we live on will double its revenge.

5 Lions can invoke their retractable claws when they're struggling with animals, and they can _____ the skin with those claws.

6 The edge of the dark circle at the center is the point at which the gas enters the black hole, which is an object that has such a large gravitational pull, not even light can _____.

7 The eco-friendly transportation program is _____ more positive than negative, because it is a good way to decrease our carbon footprint.

8 With the deterioration of the ecological environment, many _____, difficult to spot and little-known species are highly endangered.

9 It was cold yesterday and my grandfather was _____ under a black blanket.

10 The research claims that a(n) _____ lifestyle is as bad for health as smoking and obesity because of the dangers posed by inactivity.

Translation

Part A   Translate the following paragraph into Chinese.
Because all the mass of the black hole is concentrated into an infinitely tiny point, the differences in gravity are extreme. You are stretched head to toe in an aptly named process known as spaghettification. And what's more, you're squeezed along your midsection. This squeezing operates on the beams of light surrounding you as well, concentrating the infalling light into a bright band about your waist.

Part B   Translate the following paragraph into English.
今年，一组研究人员终于在实验室实验中发现了这种令人难以捉摸的霍金辐射（Hawking radiation）。该团队用一股极冷的气流创造了一个"瀑布"，用以模拟黑洞的事件视界。注入气体中的量子声波（quantum sound waves）如果被插入附近的"流"中，就可能从瀑布中流走，但瀑布本身的声波却被无情的气流困住了。这些逃逸的声波可被视为逃避黑洞引力的光粒子，这表明霍金的理论是正确的。

Write a summary of the text in 120 words.

Directions: A summary should be written in your own words. It should contain only the ideas of the original text. Do not insert any of your own opinions, interpretations, deductions or comments into the summary.

IV Read the passage, select one word for each blank from a list of choices in the bank, and change the form when necessary. You may not use any of the words in the bank more than once.

simulate	particle	field	model	perplex
momentum	universe	cosmic	spread	ignite

The Big Bang Theory, which describes 1) _____ inflation, remains the most widely supported explanation of how our universe began, yet scientists are still 2) _____ by how these wholly different periods of expansion are connected. To solve this cosmic conundrum, a team of researchers at Kenyon College, the Massachusetts Institute of Technology (MIT) and the Netherlands' Leiden University 3) _____ the critical transition between cosmic inflation and the Big Bang – a period they call "reheating".

When the universe expanded in a flash of a second during cosmic inflation, all the existing matter was 4) _____ out, leaving the universe a cold and empty place, devoid of the hot soup of particles needed to 5) _____ the Big Bang. During the reheating period, the energy propelling inflation is believed to decay into 6) _____, said Rachel Nguyen, a doctoral student in physics at the University of Illinois and lead author of the study.

"Once those particles are produced, they bounce around and knock into each other, transferring 7) _____ and energy," Nguyen said. "And that's what thermalizes and reheats the universe to set the initial conditions for the Big Bang." In their 8) _____, Nguyen and her colleagues simulated the behavior of exotic forms of matter called inflatons. Scientists think these hypothetical particles, similar in nature to the Higgs boson, created the energy 9) _____ that drove cosmic inflation. Their model showed that, under the right conditions, the energy of the inflatons could be redistributed efficiently to create the diversity of particles needed to reheat the 10) _____.

V Critical thinking

Work in groups and discuss the following questions.

Some people say that space exploration helps to address fundamental questions about our place in the universe and the history of our solar system. Through addressing the challenges related to human space exploration we develop technologies, create new industries, and help to foster a peaceful connection with other nations. Curiosity and exploration are vital to the human spirit and accepting the challenge of going deeper into space will invite the people of the world today and the generations of tomorrow to join the exciting journey to outer space.

Some argue that to study black holes is to satisfy humans' curiosity. Others say that black holes have nothing to do with our daily life. As far as you are concerned, how would you evaluate the significance of studies on black holes? What adjectives would you use to describe a black hole: mysterious, powerful or overwhelming? In your mind, what makes a black hole such a special cosmic body?

⊙ Further reading

The Kuiper Belt: Objects at the edge of the solar system

1 Beyond the gas giant Neptune lies a region of space filled with icy bodies. Known as the Kuiper Belt, this chilly expanse holds trillions of objects – remnants of the early solar system.

2 In 1943, astronomer Kenneth Edgeworth suggested comets and larger bodies might exist beyond Neptune. And in 1951, astronomer Gerard Kuiper predicted the existence of a belt of icy objects at the far edge of the solar system. Today, the rings predicted by the pair are known as the Kuiper Belt or the Edgeworth-Kuiper Belt.

3 Despite its massive size, the Kuiper Belt wasn't discovered until 1992 by astronomers Dave Jewitt and Jane Luu. According to NASA, the pair had been "doggedly scanning the heavens in search of dim objects beyond Neptune" since 1987. They dubbed the first object they spotted "Smiley" but it was later cataloged as "1992 QB1".

4 Since then, astronomers have discovered several intriguing Kuiper Belt objects (KBO) and potential planets within the region. NASA's New Horizons mission continues to uncover previously hidden planets and objects, helping scientists learn more about this unique solar system relic.

Kuiper Belt formation

5 When the solar system formed, much of the gas, dust and rocks pulled together to form the sun and planets. The planets then swept most of the remaining debris into the sun or out of the solar system. But objects at the edge of the solar system were far enough away to avoid the gravitational tugs of the much larger planets like Jupiter, and so managed to stay in their place as they slowly orbited the sun. The Kuiper Belt and its compatriot, the more distant and spherical Oort Cloud, contain the leftover remnants from the beginning of the solar system and can provide valuable insights into its birth.

6   According to the Nice model – one of the proposed models of solar system formation – the Kuiper Belt may have formed closer to the sun, near where Neptune now orbits. In this model, the planets engaged in an elaborate dance, with Neptune and Uranus changing places and moving outward, away from the sun. As the planets moved farther away from the sun, their gravity may have carried many of the Kuiper Belt objects with them, shepherding the tiny objects ahead as the ice giants migrated. As a result, many of the Kuiper Belt objects were moved from the region they were created into the colder part of the solar system.

7   The most crowded section of the Kuiper Belt is between 42 and 48 times Earth's distance from the sun. The orbit of objects in this region remains stable for the most part, although some objects occasionally have their course changed slightly when they drift.

8   Scientists estimate that thousands of bodies more than 100 km in diameter travel around the sun within this belt, along with trillions of smaller objects, many of which are short-period comets. The region also contains several dwarf planets – round worlds too large to be considered asteroids but too small to qualify as a planet.

Kuiper Belt objects

9   Pluto was the first true Kuiper Belt object to be seen, although scientists at the time didn't recognize it as such until other KBOs were discovered. Once Jewitt and Luu discovered the Kuiper Belt, astronomers soon saw that the region beyond Neptune was full of icy rocks and tiny worlds.

10 Sedna, a KBO that's about three-quarters the size of Pluto, was discovered in 2004. It is so far out from the sun that it takes about 10,500 years to make a single orbit. Sedna is about 1,770 km wide and circles the sun in an eccentric orbit that ranges between 12.9 billion km and 135 billion km.

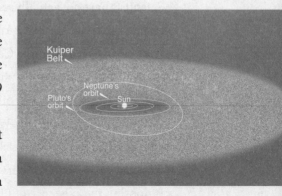

11 "The sun appears so small from that distance that you could completely block it out with the head of a pin," Mike Brown, an astronomer at the California

Institute of Technology who discovered this and several other Kuiper Belt objects, said in a statement.

12 In July 2005, astronomers discovered Eris, a KBO that's slightly smaller than Pluto. Eris orbits the sun approximately once every 580 years, traveling almost 100 times farther from the sun than Earth does. Its discovery revealed to some astronomers the problem of categorizing Pluto as a full-scale planet. According to the 2006 definition of the International Astronomical Union (IAU), a planet must be large enough to clear its neighborhood of debris. Pluto and Eris, surrounded by the Kuiper Belt, had clearly failed to do so. As a result, in 2006, Pluto, Eris, and the largest asteroid, Ceres, were reclassified by the IAU as dwarf planets. Two more dwarf planets, Haumea and Makemake, were discovered in the Kuiper Belt in 2008.

13 Astronomers are now reconsidering Haumea's status as a dwarf planet. In 2017, when the object passed between Earth and a bright star, scientists realized it is more elongated than round. Roundness is one of the criteria of a dwarf planet, according to the IAU's definition. Haumea's elongated shape could be a result of its rapid spin; a day on the object only lasts about four hours.

14 "I don't know if this will change the definition of a dwarf planet," said Pablo Santos-Sanz, an astrophysicist at the Institute of Astrophysics of Andalusia in Spain. "I think probably yes, but probably it will take time."

Planet Nine

15 Planet Nine is a hypothetical world thought to orbit the sun at a distance that is about 600 times farther from the sun than Earth's orbit, and about 20 times farther out than the orbit of Neptune. (The orbit of Neptune is 4.3 billion km from the sun at its closest point.)

16 Scientists have not actually seen Planet Nine. Its existence was inferred by gravitational effects observed on other objects in the Kuiper Belt. Scientists Mike Brown and Konstantin Batygin at the California Institute of Technology described the evidence for Planet Nine in a study published in the *Astronomical Journal* in 2016.

17 If there is another world out there, astronomers Scott Sheppard, of the Carnegie Institution for Science in Washington, D.C., and Chadwick Trujillo, of Northern Arizona University, are likely to find it soon. The pair has spent the last six years working on the deepest survey of faint objects at the edge of the solar system, after proposing the existence of Planet X, a small dwarf planet beyond Pluto, in 2014.

18 So far, Sheppard and Trujillo have found 62 distant objects, which make up about 80 percent of all of those at the edge of the system. Last year, the two discovered the dwarf planet 2015 TG387, nicknamed "The Goblin", and the most distant KBO ever reported, 2018 VG18, nicknamed "FarOut". In February 2019, Sheppard unofficially announced the discovery of an even more distant object, informally known as "FarFarOut".

19 "These distant objects are like breadcrumbs leading us to Planet X," Sheppard said in a statement. "The more of them we can find, the better we can understand the outer solar system and the possible planet that we think is shaping their orbits – a discovery that would redefine our knowledge of the solar system's evolution."

New words

Neptune /ˈneptjuːn/ *n.* the planet in the solar system that is eighth in order of distance from the sun 海王星

remnant /ˈremnənt/ *n.* a part of something that is left after the other parts have been used, destroyed or eaten 残余部分, 剩余部分

doggedly /ˈdɒɡɪdli/ *adv.* in a very determined way, even if something is very difficult 坚持不懈地

dub /dʌb/ *v.* to give sb./sth. a particular name, often in a humorous or critical way 把…戏称为; 给…起绰号

smiley /ˈsmaɪli/ *n.* a sign that looks like a face to show that you are happy or pleased about sth. 笑容符

catalog /ˈkætəlɒɡ/ *v.* to record sth. in a list 把…编入目录

intriguing /ɪnˈtriːɡɪŋ/ *adj.* very interesting because of being unusual or not having an obvious answer 非常有趣的; 引人入胜的; 神秘的

relic /ˈrelɪk/ *n.* an old object or custom that reminds people of the past or that has lived on from a past time 遗迹; 遗俗

compatriot /kəmˈpætriət/ *n.* someone who was born in, or is a citizen of, the same country as someone else 同胞, 同国人

spherical /ˈsferɪkəl/ *adj.* having the shape of a sphere 球形的, 球状的

shepherd /ˈʃepəd/ *v.* to guide or lead a group of people somewhere, making sure they go where you want them to go 带领, 引导

Pluto /ˈpluːtəʊ/ *n.* a dwarf planet in the Kuiper Belt, a ring of bodies beyond the orbit of Neptune 冥王星

eccentric /ɪkˈsentrɪk/ *adj.* not having the same center point 不同圆心的

elongated /ˈiːlɒŋɡeɪtɪd/ *adj.* long and thin, often in a way that is not normal 拉长的, 偏长的

goblin /ˈɡɒblɪn/ *n.* a small ugly creature in children's stories that likes to trick people or cause trouble (童话中的) 小妖精

breadcrumbs /ˈbredkrʌmz/ *n.* [plural 复数] tiny pieces of bread 面包屑

Phrases and expressions

at the edge of 在…的边缘
pull together 聚集; 齐心协力

block out 挡住, 遮住

Proper nouns

Kuiper Belt 柯伊伯带

Edgeworth–Kuiper Belt 埃奇沃思－柯
伊伯带

Oort Cloud 奥尔特云

Sedna 塞德娜星

California Institute of Technology 加
州理工学院

Eris 厄里斯星，阋神星

International Astronomical Union
(IAU) 国际天文学联合会

Ceres 刻瑞斯星，谷神星

Haumea 妊神星

Makemake 鸟神星

Institute of Astrophysics of Andalusia
西班牙安达卢西亚天体物理研究所

Planet Nine 第九行星

Astronomical Journal《天文学报》

Carnegie Institution for Science 卡内基
科学研究所

Northern Arizona University 北亚利桑
那大学

Technical terms

Kuiper Belt object (KBO) 柯伊伯带天体
New Horizons mission "新视野号" 任务
gravitational tug 引力牵引

Nice model 尼斯模型
dwarf planet 矮行星

Exercises

For each of the following unfinished statements or questions, choose the most appropriate answer from A, B, C, or D according to the text.

1 _____ first discovered the Kuiper Belt.
   A. Kenneth Edgeworth   B. Gerard Kuiper
   C. Dave Jewitt and Jane Luu D. Mike Brown

2 According to the scientists' estimation, the bodies that travel around the sun within the Kuiper Belt include _____.
   A. short-period comets   B. dwarf planets
   C. asteroids   D. A and B

3   Which one is NOT a Kuiper Belt object?

   A. Pluto.   B. Eris.

   C. Sedna.   D. Ceres.

4   Why do the astronomers suspect that Haumea may not be a dwarf planet?

   A. Because it is not round.   B. Because it is not bright.

   C. Because it is not near the sun.   D. Because it has an eccentric orbit.

5   Planet Nine was found by _____.

   A. a telescopic observation   B. nobody but scientists' deduction

   C. an astronomer   D. a picture from a spacecraft

Questions for discussion

1   What is the Kuiper Belt and where is it?

2   What is the significance of the study of the Kuiper Belt and Oort Cloud?

3   Why is "the head of pin" mentioned in Para. 11?

4   What is the significance of the discovery of Eris?

5   Why were Pluto and Eris reclassified by the IAU as dwarf planets?

⊙ Practical writing

How to make a science and technology investigation report

A science and technology investigation report refers to a document about the investigation of a job, event, problem, etc. It systematically sorts out, analyzes and studies the materials collected during the investigation and presents the whole analysis in an objective way. The science and technology investigation report features in realism, pertinence and logic.

Science and technology investigation reports can be mainly divided into three types: situation investigation reports, typical experience investigation reports and problem investigation reports.

A science and technology investigation report often consists of two parts: the heading and body. Headings can be written in two ways. One is a standardized format and the basic ones are "An investigation report about ...", "A survey report on ...", etc. The other is the freestyle title, including the questioning type and the type of combining the heading and the subtitle. The questioning type will pose a question in the heading such as "Why ...". The type of combining the heading and the subtitle is a little special. Its heading just focuses on the main conclusion of the investigation or poses a key question of the investigation. Its subtitle mainly poses its research subject, research field and research question.

The body of a science and technology investigation report includes three parts: preface, main body and ending. The preface is to specify the cause, purpose, time, place, object, scope, process and method of the investigation, the central question or basic conclusion. The main body is the most important part of the survey report, which specifies the basic information, practices and experiences of the survey, as well as the specific understanding, views and conclusions obtained from the analysis of the materials of the survey. The ending of the writing can be diversified as following: It puts forward the solution to the problem, countermeasures or suggestions for the next step to improve the work; it summarizes the main points of the full text and further deepens the theme; it asks questions that provoke further thought; it looks forward to the future and gives encouragement and call.

Exercise

Write an investigation report on aviation science and technology museums, including the preface, main body and ending. Please investigate the necessity of building aviation science and technology museums, the problem(s) of their development, and their developing tendency.

Unit **7**
Eco-friendly flying

Learning objectives

Upon completion of this unit, you will be able to:

- identify the measures to reduce the carbon footprint of flying;
- analyze the process in which greenhouse gases cause climate change;
- design feasible and workable plans for green travel in your daily life;
- write an article review.

◉ **Lead-in**

With the development of civil aviation, air travel has become an important part of people's lives. However, have you ever considered the impact that flying has on the environment? Has it occurred to you that people could make flying less carbon-intensive? Fortunately, since the aviation industry has been confronted with the rise of the flight shaming movement, some travelers are shunning air travel and becoming more aware of their environmental impact. What's more, some tips, such as flying less and carrying less stuff, are advocated to make flying more carbon-efficient.

⊙ Intensive reading

Trying to make flying less carbon-intensive

1　Here's something to ponder as you think about making your New Year's resolutions: There's something you could skip just one or two times a year that could reduce your carbon footprint by as much as 10 to 20 percent. It's something that's kind of a hassle anyway and costs a pretty penny to boot. To be on your way of having the carbon footprint of a bicycle-riding, plastic-recycling European, all you have to do is to cut one or two flights a year out of your life.

2　If you do fly, there are also tips for making it more carbon-efficient. It's a choice more and more people are making, in part because air travel puts a lot of climate-warming carbon dioxide into the atmosphere. When Karyn Hunt began planning a trip to Disneyland, she faced a choice. Her family of five could catch a flight from their San Francisco-area home or they could drive. Their first thought was to fly from San Francisco International Airport. But over the past few years Hunt and her husband have begun making small steps to minimize the amount of climate-change-causing carbon they add to the atmosphere. Flying less is one of those changes. "When I'm out on the freeway I can see one plane after another launching from SFO, and I think 'This just doesn't make sense. We're traveling too much,'" she said. So instead of hopping on a flight, they piled into their plug-in hybrid car and drove the 380 miles to Anaheim, California. "It was cramped and a little uncomfortable, but it felt like the right decision," she said.

3　For when you do fly, buying carbon offsets to compensate for the carbon dioxide the flight produces is another option. That was the choice of Bishop Marc Andrus, of the Episcopal Diocese of California, and he measured his carbon footprint using an app, SustainIslandHome.org. "It was not good. I fly a lot," he said. All but one of his flights in the previous year had been for work, for which he needed to be physically present. Instead, he chose to buy carbon offsets to reduce CO_2 emissions in another area, equivalent to what his flights were producing. "The money went to preserve prairie land in the United States and old-growth forests in Peru. It is less than the cost of a checked bag per flight! People don't realize how reasonable it is," he said. If lowering your carbon footprint is on your possible list of New Year's resolutions, here are some tips.

4 **Fly less.** Flying takes a lot of energy, which means releasing a lot of carbon dioxide into the atmosphere. There's just no way around it: Creating the thrust necessary to push a 130,000 pound airplane 35,000 feet above the Earth, keep it there for a couple of hours and then bring it down safely takes a lot of jet fuel. "It really does matter. If there's one thing a single person can do with maximum effect, it's thinking about their flights," said Dietrich Brockhagen, executive director of Atmosfair, a German non-profit that focuses on flight emissions.

5 How often you fly also matters. The average per capita emission of carbon for the Americans is about 16 metric tons, said Stefan Gössling, an economics professor at Sweden's Linnaeus University and co-editor of the book, *Climate Change and Aviation: Issues, Challenges and Solutions.* One flight from the West to the East coast across the US produces at a minimum one metric ton of carbon dioxide. If all climate-change-causing emissions are included, one flight from the United States to Asia or

from Asia to Europe can produce as much as five metric tons of carbon equivalent emissions, which includes both carbon dioxide and other greenhouse gases, he said. To put that in perspective, five metric tons is the average amount of carbon dioxide produced by every human each year on the planet.

6　Burning jet fuel releases greenhouse gases such as carbon dioxide into the Earth's atmosphere and oceans. Greenhouse gases block heat escaping from the atmosphere, causing temperatures to rise just like in a greenhouse. "About eight percent of global greenhouse gas emissions are from traveling," including air travel, hotels, food and sundries, said Arunima Malik, a researcher at the University of Sydney in Australia who studies the carbon footprint of tourism.

7　Humans have increased the Earth's atmospheric carbon dioxide concentration by more than a third since the Industrial Revolution began, according to NASA. The extra carbon dioxide has caused temperatures to rise to levels that cannot be explained by natural factors, scientists report. During the 20th century the Earth's average temperature increased about two degrees Fahrenheit, data from the National Oceanic and Atmospheric Administration shows.

8　**Nonstop is best.** Flying nonstop is much more energy-efficient than flying multiple hops. That's because takeoffs and climb require significantly more energy than cruising at an altitude – as much as 75 percent of fuel usage on a flight of 430 miles, said Dan Rutherford, director of aviation programs at the International Council on Clean Transportation. So the fewer times you have to take off per trip, the better. "The optimal length of a flight is 3,000 miles, so basically across the United States," he said. That's because really long-haul flights, such as from San Francisco to Beijing, require planes to carry extra fuel, making them heavier and less efficient.

9　**No short hops.** From a scientific viewpoint, everything that's below 600 miles, taking a train, a bus or driving is much more efficient, especially if you have more than one person in the car, said Gössling. Buses are becoming more popular in the United States. Millennials and post-millennials especially have embraced Greyhound and new bus carriers such as Megabus and BoltBus that offer cheap tickets, WiFi and mobile booking apps. They have been expanding rapidly since 2014, according to researchers at DePaul University in Chicago.

10 **Stay longer.** One option is taking one long vacation rather than two short ones, or flying to one destination and staying put. "You gain a more intimate relation to the destination, so if you really want to get to know the people there, stay longer. In one week you can't accomplish much but in two weeks you can," said Brockhagen.

11 **Don't fly business.** The amount of energy required to fly a plane is divided among the people being flown. The more people, the more energy-efficient. Business class and first class seat fewer people, so they're less efficient. Depending on the size of the area for each seat, business class is usually between two or three times as energy-intensive as economy class. First class, especially when it includes lie-flat beds, can be as much as four times as energy-intensive, said Gössling. And it doesn't get you off the hook to simply say the seats would be there whether you sit in them or not. Buying business class seats encourages airlines to create more in their planes, making them overall less efficient per passenger. "You're encouraging the airlines to install more business class seating so it's a net negative," said Rutherford.

12 **Carry less stuff.** The more your baggage weighs, the more the plane has to carry and the less energy-efficient it is. So pack light.

13 **Consider carbon offsets.** These are programs run by non-profits that allow you to purchase a carbon offset equivalent to how much carbon your flight costs. For example, Atmosfair has a program that subsidizes the cost of fuel-efficient cookstoves sold in Africa that decrease the amount of wood and other carbon-emitting fuel people must burn to cook. "You're effectively paying money into something that seeks to save energy and emissions elsewhere," said Gössling. Many US airlines also partner with carbon offset programs, including Delta, JetBlue and United, though you have to search their websites to find the information.

New words

hassle /ˈhæsəl/ *n.* a situation that is annoying because it involves doing sth. difficult or complicated that needs a lot of effort 麻烦; 困境

cramped /kræmpt/ *adj.* not having enough space 狭窄的, 狭小的

offset /ˈɒfset/ *n.* a compensating equivalent that serves to counterbalance or to compensate for sth. else 抵消; 补偿

episcopal /ɪˈpɪskəpəl/ *adj.* being connected with a bishop or bishops 主教的; 主教制的

diocese /ˈdaɪəsɪs/ *n.* a district for which a bishop is responsible 教区, 主教辖区

emission /ɪˈmɪʃən/ *n.* a gas, etc. that is sent out into the air 排放物

thrust /θrʌst/ *n.* the force of an engine that makes a car, train, or plane move forward 推力, 推进力

non-profit /ˌnɒn ˈprɒfɪt/ *n.* an organization using the money it earns to help people instead of making a profit 非营利组织, 非营利机构

millennial /mɪˈleniəl/ *n.* a person who was born around the time of the millennium, that is around the year 2000 千禧一代

subsidize /ˈsʌbsɪdaɪz/ *v.* to pay some of the cost of goods or services so that they can be sold to other people at a lower price 给…津贴 (或补贴)

Phrases and expressions

cost a pretty penny 花费很多钱

no way around 没办法

stay put 留在原地

get sb. off the hook 帮某人摆脱困境

Proper nouns

San Francisco International Airport (SFO) 旧金山国际机场

Episcopal Diocese of California 圣公会加州教区

Industrial Revolution 工业革命

National Oceanic and Atmospheric Administration 美国国家海洋和大气管理局

Greyhound 灰狗巴士

Megabus 超级巴士

BoltBus 闪电巴士

Delta 美国达美航空公司

JetBlue 美国捷蓝航空公司

United 美国联合航空公司

Technical terms

carbon-intensive 碳密集型的   plug-in hybrid car 插电式混合动力车
carbon footprint 碳足迹   greenhouse gas 温室气体

Background information

Atmosfair (德国的一个非营利环保组织)

Atmosfair is an independent German non-profit organization which offers offsets for greenhouse gases emitted by aircraft, cruise ships, long-distance coaches, and events. The organization develops and finances small-scale energy efficiency and renewable energy projects in developing countries, which lead to reduced carbon emissions.

International Council on Clean Transportation 国际清洁运输理事会

The International Council on Clean Transportation (ICCT) is an independent non-profit organization to provide first-rate, unbiased research and technical and scientific analysis to guide the environmental regulation of transportation technologies and fuels. Since the ICCT's founding, its mission has centered on improving the environmental performance and energy efficiency of road, marine, and air transportation in ways that benefit public health and mitigate climate change.

Difficult sentences

1 **There's something you could skip just one or two times a year that could reduce your carbon footprint by as much as 10 to 20 percent. (Para. 1)**

By not doing something one or two times a year, you could lower your carbon footprint by as much as 10 to 20 percent.

2 **To be on your way of having the carbon footprint of a bicycle-riding, plastic-recycling European, all you have to do is to cut one or two flights a year out of your life. (Para. 1)**

To become an environment-friendly European who rides a bicycle and recycles plastic items, you just need to make some changes in your life by cutting one or two flights a year.

3　**It's a choice more and more people are making, in part because air travel puts a lot of climate-warming carbon dioxide into the atmosphere. (Para. 2)**

More and more people are trying to cut flights in their life partly because flying releases greenhouse gases and increases temperatures.

4　**But over the past few years Hunt and her husband have begun making small steps to minimize the amount of climate-change-causing carbon they add to the atmosphere. (Para. 2)**

However, in recent years, Mr. and Mrs. Hunt have been making some changes to reduce as many carbon emissions as possible.

5　**For when you do fly, buying carbon offsets to compensate for the carbon dioxide the flight produces is another option. (Para. 3)**

If you fly, something else you can do is to buy carbon offsets to make up for the carbon dioxide your flight produces.

6　**Instead, he chose to buy carbon offsets to reduce CO_2 emissions in another area, equivalent to what his flights were producing. (Para. 3)**

In order to counterbalance the carbon dioxide produced by his flights, he bought carbon offsets equivalent to what has been produced.

7　**There's just no way around it: Creating the thrust necessary to push a 130,000 pound airplane 35,000 feet above the Earth, keep it there for a couple of hours and then bring it down safely takes a lot of jet fuel. (Para. 4)**

There's nothing we can do because it takes a lot of jet fuel to fly a plane from its takeoff to landing.

8　**If all climate-change-causing emissions are included, one flight from the United States to Asia or from Asia to Europe can produce as much as five metric tons of carbon equivalent emissions, which includes both carbon dioxide and other greenhouse gases, he said. (Para. 5)**

If all the emissions that cause climate change are calculated, the carbon equivalent emissions of both carbon dioxide and other greenhouse gasses produced from one flight from the United States to Asia or from Asia to Europe can be as much as five metric tons.

9 **These are programs run by non-profits that allow you to purchase a carbon offset equivalent to how much carbon your flight costs. (Para. 13)**

Organizations not aiming to make money provide carbon offset programs for you to counterbalance the carbon produced by your flight.

10 **For example, Atmosfair has a program that subsidizes the cost of fuel-efficient cookstoves sold in Africa that decrease the amount of wood and other carbon-emitting fuel people must burn to cook. (Para. 13)**

For example, Atmosfair pays part of the cost of energy-efficient cookstoves sold in Africa that reduce the amount of wood and fuel required so that less carbon will be released.

Exercises

| Fill in the blanks with the words and expressions given below, and change the forms when necessary.

> intensive   cruise   compensate   offset   millennial
>
> emission   hybrid   subsidize   non-profit   hassle

1 A(n) _____ vehicle is one that is powered by both electricity and a traditional fuel, gasoline.

2 This foundation is a(n) _____ whose objective consists in promoting the success of emerging agent-based technology.

3 California promises to use 60 percent renewable energy sources for its power generation by 2030 and all zero _____ sources by 2045.

4 In some areas, modern _____ farming is giving way to the reintroduction of traditional methods.

5 That lack of curiosity produces a relative lack of knowledge, and the lack of knowledge is difficult if not impossible to _____ for later on.

6 At a typical _____ altitude, up to eight pounds of pressure are pushing against every square inch of the body of the aircraft.

7 He raises the question as to whether the fine arts should be _____ by public funds.

8 _____ have grown up with the internet and can't imagine a world without it.

9 It was such a(n) _____ trying to get my bank account changed that I nearly gave up.

10 A carbon _____ is a reduction or removal of emissions of carbon dioxide or other greenhouse gases made in order to compensate for emissions made elsewhere.

|| Translation

Part A   Translate the following paragraph into Chinese.

Flying takes a lot of energy, which means releasing a lot of carbon dioxide into the atmosphere. There's just no way around it: Creating the thrust necessary to push a 130,000 pound airplane 35,000 feet above the Earth, keep it there for a couple of hours and then bring it down safely takes a lot of jet fuel. "It really does matter. If there's one thing a single person can do with maximum effect, it's thinking about their flights," said Dietrich Brockhagen, executive director of Atmosfair, a German non-profit that focuses on flight emissions.

Part B   Translate the following paragraph into English.

2019年12月，在日内瓦举行的国际航空运输协会（International Air Transport Association）"全球媒体日"会议上，协会理事长兼首席执行官亚历山大·德·朱尼亚克（Alexandre de Juniac）强调："飞行不是敌人，碳排放才是问题。"他还特别指出，早在飞行耻辱运动（flight shaming movement）开展之前，航空业就已提出必须"认真应对气候变化"。

||| Write a summary of the text in 120 words.

Directions: A summary should be written in your own words. It should contain only the ideas of the original text. Do not insert any of your own opinions, interpretations, deductions or comments into the summary.

|V Read the passage, select one word for each blank from a list of choices in the bank, and change the form when necessary. You may not use any of the words in the bank more than once.

generate	flight	greenhouse	cruise	aviation
footprint	emission	atmosphere	travel	distance

Take one round-trip flight between New York and California, and you've generated about 20 percent of the 1) ＿＿＿＿＿ gases that your car emits over an entire year. If you are like many people, flying may be a large portion of your carbon footprint. Over all, the 2) ＿＿＿＿＿ industry accounts for 11 percent of all transportation-related 3) ＿＿＿＿＿ in the United States. According to some estimates, about 20,000 planes are in use around the world, serving three billion passengers annually. By 2040, more than 50,000 planes could be in service, and they are expected to fly more often. If you're flying, you're adding a significant amount of planet-warming gases to the 4) ＿＿＿＿＿ – there's no way around it. But there are some ways to make your airplane 5) ＿＿＿＿＿ a little bit greener.

The most effective way to reduce your carbon 6) ＿＿＿＿＿ is to fly less often. If everyone took fewer 7) ＿＿＿＿＿, airline companies wouldn't burn as much jet fuel. According to the World Bank, the average American 8) ＿＿＿＿＿ about 16.4 metric tons of carbon dioxide in 2013; according to some calculations, a round-trip flight from New York to San Francisco emits about 0.9 metric tons of carbon dioxide per person. For an American, that represents about one-eighteenth of your carbon emissions for the year. Should you drive instead? The longer the 9) ＿＿＿＿＿, the more efficient flying becomes, because 10) ＿＿＿＿＿ requires less fuel than other stages of flight.

Ⅴ Critical thinking

Work in groups and discuss the following questions.
Global warming may get worse in the coming years even if humans immediately stopped producing carbon dioxide. So "green" is becoming a preference in many people's minds. Read the text again, and think about the following questions. What is eco-friendly transportation? What benefits would eco-friendly transportation bring on human society as well as on the environment? What measures can we take to reduce carbon emissions?

⊙ **Further reading**

Could eco-friendly flying be on the horizon?

1　Air travel is one of the fastest-growing sources of carbon emissions. But could green tech and optimization mean that one day, we'll be able to hop on a plane with a guilt-free conscience? A new hashtag from Sweden has captured a peculiar 21st century angst: flygskam – or flying shame. That's the gnawing awareness that you're consuming more than your fair share of the carbon budget.

2　To keep within the two degrees of global warming, scientists have calculated that we each have a personal allowance of about two tons of CO_2 per year. In Germany, per capita emissions are about 4.5 times that; in the United States around eight, which means hopping on a plane for a jolly time does feel pretty shameful. Björn Ferry, a former Olympic athlete, Swedish sports commentator and now early advocate of the flygskam movement, set himself the goal of living completely fossil-fuel free by 2025. That meant buying more second-hand, eating less meat and growing his own veggies. But perhaps the biggest change was how he travels. "I decided to stop flying almost two years ago," he said. "It's impossible to be fossil-free with flying at the moment."

Eco-planes of the future?

3　Battery-powered planes are still in the early stages of development, and experts don't expect them to be rolled out on a large scale within the next two decades. But it is already possible to manufacture aircraft fuel from renewable power. Water is electrolyzed to generate hydrogen, CO_2 from the atmosphere is added and – presto! – you have synthetic kerosene that fuels planes just fine. Power-to-liquid (PtL as it's called), is preferable to biofuels which come with their own environmental costs, primarily because of the huge amount of land needed to grow the raw material. But so far, PtL fuels are pricy. Experts say a price on carbon could change the math, allowing producers in countries that generate large volumes of affordable sun, wind and hydropower to compete against dirty old fossil fuels. Norway, which gets the vast majority of its electricity from hydropower, already has a large-scale facility that's set to start producing the stuff on an industrial scale in 2020.

4   So is climate-neutral air travel within reach? Well, aside from an effective carbon price being nowhere on the horizon, the answer is ... it's complicated because CO_2 emissions aren't the only factor behind planes heating up the planet.

The real contrail conspiracy

5   At high altitudes, contrails – that is, clouds of ice crystals and ozone from air traffic – have a greenhouse effect as well. In fact, this currently causes about twice as much climate impact as the CO_2 from flying itself. And that effect is just as bad regardless of whether the fuel is fossil or synthetic.

6   Clouds from aircraft reflect the sun's rays back into space. But that effect can be reduced – "or even reversed," according to Stefanie Meilinger, a professor of sustainable technology at the Bonn-Rhein-Sieg University of Applied Sciences in Germany – by "optimizing" routes. That would include measures such as flying at lower altitudes during certain times of day, or at night instead – and planning shorter routes that consume less fuel.

7   Working with the German Aerospace Center (DLR) and National Meteorological Service, Meilinger is developing a system to optimize routes for Lufthansa Systems. Urban Weisshaar of Lufthansa Systems says optimized routes could even be cheaper,

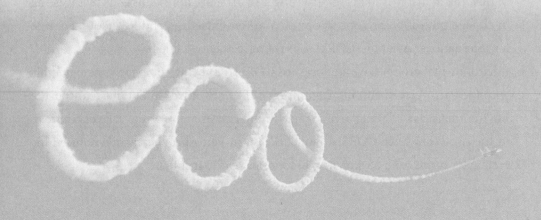

and the most harmful routes could even be closed, if airlines choosing them were rewarded with lower taxes or fees. Then again, cutting prices and offering further tax breaks seems a little counterintuitive in an industry that already benefits globally from abundant subsidies and tax breaks.

Who pays the price?

8   Air tickets are crazy light on the wallet – especially when you think of the heavy price the environment pays. In fact, because air fuel isn't taxed at all, flying is often the cheapest option for journeys that could just as well be taken with fewer emissions. If the costs of environmental damage were reflected in ticket prices – for example through effective carbon pricing or heavy taxation – they would skyrocket, making buses and trains look a whole lot more attractive.

9   The Stay Grounded network says such cleaner options need investment too, to improve services and help them compete against quick and dirty air travel. Stay Grounded is also calling for a global halt to the construction of new airports, policies to promote more localized economies, and a ban on advertising such as many countries have for tobacco products. Arne Fellermann of Friends of the Earth Germany, which is a member of the Stay Grounded network, would like to see a national ban on short-haul flights in his country, too. So far, the aviation industry has shown little sign of being ready to act on its own accord.

Hollow promises

10   In 2016, the United Nations aviation agency ICAO launched a much-anticipated industry-wide agreement aimed at making the sector greener. Yet it contained no measures for replacing fossil fuels or reducing other greenhouse gas emissions. Perhaps worst of all, it also assumed the industry would continue to grow unchecked. Rather, the industry is to begin offsetting a small share of its emissions by investing in climate protection projects under the Carbon Offsetting and Reduction Scheme for International Aviation (CORSIA, as the agreement is called).

11   Environmental groups have admonished CORSIA as being completely incompatible with the Paris Agreement. Bill Hemmings of Brussel-based think tank Transport & Environment says it "won't save a single drop of kerosene." For the sake of contrast, in the two years since CORSIA launched, Björn Ferry has meanwhile reduced his carbon footprint from a whopping 15 tons per year to just four.

New words

gnawing /ˈnɔːɪŋ/ *adj.* making you feel worried over a long period of time（长时间）折磨人的，令人痛苦的，使人苦恼的

commentator /ˈkɒmənteɪtə/ *n.* a person who is an expert on a particular subject and talks or writes about it on television or radio, or in a newspaper（电视台、电台或报刊的）评论员

electrolyze /ɪˈlektrəlaɪz/ *v.* to subject to or treat by electrolysis 电解

synthetic /sɪnˈθetɪk/ *adj.* produced by combining different artificial substances, rather than being naturally produced 合成的，人造的

kerosene /ˈkerəsiːn/ *n.* a clear oil that is burnt to provide heat or light 煤油

contrail /ˈkɒntreɪl/ *n.* a line of white steam made in the sky by a plane（飞机在空中飞行留下的）凝结尾流，拉烟

conspiracy /kənˈspɪrəsi/ *n.* a secret plan by a group of people to do sth. harmful or illegal 密谋，阴谋

crystal /ˈkrɪstl/ *n.* a small piece of a substance with many even sides, that is formed naturally when the substance becomes solid 结晶；晶体

ozone /ˈəʊzəʊn/ *n.* a poisonous gas with a strong smell that is a form of oxygen 臭氧

Phrases and expressions

roll out 在市场上推出

on one's own accord 出于自愿，主动地

hollow promise 空洞的承诺

Proper nouns

German Aerospace Center (DLR) 德国航空航天中心

National Meteorological Service 德国国家气象局

Stay Grounded "脚踏实地"网络组织

ICAO (International Civil Aviation Organization) 国际民用航空组织（简称"民航组织"）

Carbon Offsetting and Reduction Scheme for International Aviation (CORSIA) 国际航空碳抵消和减排计划

Paris Agreement《巴黎协定》

Technical terms

battery-powered plane 电池动力飞机   power-to-liquid (PtL) 电转液

Exercises

| For each of the following unfinished statements or questions, choose the most appropriate answer from A, B, C, or D according to the text.

1   It is assumed that we each have a personal allowance of about two tons of CO_2 per year in order to _____.

A. keep within the two degrees of global warming

B. save money

C. reduce waste

D. all of the above

2   Which is the author's preference concerning the following sources of airplane power?

A. PtL>battery>biofuels

B. PtL>biofuels>battery

C. battery>PtL>biofuels

D. biofuels>PtL>battery

3   Will Meilinger's system of optimized routes be practical, and why?

A. No, because they will not be cheaper.

B. Yes, because they can reduce carbon emissions.

C. No, because they will not get the government's subsidies.

D. Yes, if airlines were rewarded with lower taxes or fees.

4   Why are buses and trains less attractive than flying at present according to the text?

A. Because flying looks cool.

B. Because flying is cheaper.

C. Because flying is faster.

D. All of the above.

5 The agreement launched by ICAO in 2016 endeavored to make the aviation

    sector ＿＿＿＿＿＿.

    A. cheaper

    B. greener

    C. faster

    D. higher

‖ Questions for discussion

1 What is the priority in Björn Ferry's goal by 2025?

2 Why is power-to-liquid (PtL) preferable to biofuels?

3 What are the measures to reduce the effect produced by the clouds from aircraft reflecting the sun's rays back into space?

4 What does the Stay Grounded network advocate?

5 What does the author mean by "hollow promises" when referring to the agreement launched by ICAO?

⊙ Practical writing

How to write an article review

An article review is both a summary and an evaluation of another author's article. Understanding the main points and arguments of the article is essential for an accurate summation. Logical evaluation of the article's main theme, supporting arguments, and implications for further research is an important element of a review. Here are a few steps for writing an article review.

Step 1: Come up with a title.

A title should reflect the focus of your review. Respectively, the title can be interrogative, descriptive, or declarative.

Step 2: Cite the article.

Next, create a proper citation for the reviewed article and input it following the title. At this step, the most important thing to keep in mind is the style of citation. For example, an article citation in the MLA style should look as follows:

Author's last and first name. "The title of the article." Journal's title and issue (publication date): page(s). Print.

Example: Abraham John. "The World of Dreams." *Virginia Quarterly*, vol. 60, no. 2, 1991: pp. 125-167. Print.

Step 3: Identify the article.

Start your review by referring to the title and author of the article, the title of the journal, and the year of publication in the first paragraph.

For example: The article "Implementing Artificial Intelligence and Machine Learning into Advanced Qualification Programs" was written by J. R. Herr, published in the *Journal of Aviation/Aerospace Education and Research*, vol. 30, 2021.

Step 4: Write the introduction.

The introduction of an article review will have an identification sentence. It will also mention the central themes of the article with the arguments and claims of the author. You also need to state the author's thesis. Sometimes, the thesis has multiple points. The thesis may not be clearly stated in the article, so you may have to determine the thesis yourself.

Step 5: Summarize the article.

Express the main points, arguments, and findings of the article in your own words, referring to your summary for assistance. Show how the article supports its claims. Make sure to include the article's conclusions. This may be done in several paragraphs, although the length will depend on requirements established by your instructor or publisher.

Step 6: Write your critique.

Express your opinion about whether the article was a clear, thorough, and useful explanation of the subject. Present the strengths and weaknesses you have found in the publication. Highlight the knowledge that the author has contributed to the field. Also, write about any gaps and/or contradictions you have found in the article. Take a standpoint of either supporting or not supporting the author's assertions, but back up your arguments with facts and relevant theories that are pertinent to that area of knowledge.

Step7: Conclude the article review.

Summarize the main points of the article, as well as your opinions about its significance, accuracy, and clarity. If relevant, also comment on implications for further research or discussion in the field.

Exercise

Write an article review based on an article you have read recently. You should write no less than 300 words.

Unit 8
Aviation and entertainment

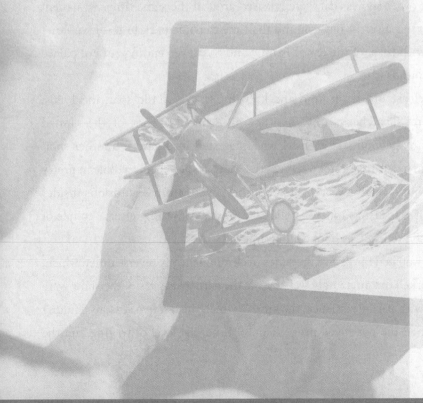

Learning objectives

Upon completion of this unit, you will be able to:

- identify how the elements of aviation are used in movies or TV series;
- appraise the potential stimulus of entertainment to the progress of aviation;
- interpret airplane myths in the movie industry in an objective way;
- write meeting minutes.

◉ Lead-in

Aviation has provided us with unprecedented opportunities for entertainment, as airplanes and spacecraft are widely applied in movies, TV series, and other popular culture. Many movies and television shows are relevant to subjects related to aviation, which boosts the technical progress in their production and sparks moviegoers' interest. In order to look at airplane myths in the movie industry sensibly, a better understanding of them is necessary.

◉ Intensive reading

Netflix's *The Crown*: A love letter to aviation

1　After capping off its fourth season, it is no secret that Netflix's *The Crown*, which follows the reign of Queen Elizabeth II, is a cinematic feat. While the acting, writing, sets, and costumes all get much acclaim, one element of the show's world-building magic that doesn't get enough attention is its near-perfect attention to detail in how it portrays aviation's intersection with the Royals' lives. In fact, I would go as far as to say that aviation is another supporting character of *The Crown*'s superb and ever-morphing cast. When watched through its current four-season run, it becomes clear that the series subtly acts as a love letter to aviation in its own right.

2　I came to watch *The Crown* via a circuitous route. I checked out the first episode half a decade ago when it premiered and, for whatever reason, found it a bit dour. In retrospect, I quit not just due to that first impression really, but more so because of distractions from other shows that were released around the same time. As it went on to win award after award I never found the time to come back to it – I wouldn't call myself a big enthusiast of the Royal Family and the never-ending aura of palace intrigue that surrounds it.

3　Then, with the release of season four, and knowing that it would span the 1980s, including Margaret Thatcher's time in office as Prime Minister of the United Kingdom and many geopolitical events I was interested in, I gave it another shot. What I quickly learned is that it not only was the show absolutely incredible in pretty much every way, but it really wasn't just about the Royal Family's notorious insider melodrama, and it was just as much a political drama as anything else. A couple of episodes into season four, I was hooked, big time.

4　Since then, I have watched the rest of the series in chronologically reverse order, which, oddly enough, worked out amazingly well. While the earlier seasons of the show were definitely stuffed with period romance, the kind of which the last two seasons couldn't compete with due to the portrayed progression of time, the level of writing and the complexity of the storytelling evolved triumphantly throughout the series back half. Some episodes are working on so many levels that you have to really take a step back to admire just how well written and edited they are. Still, the production – that marvelous production – really is worth the price of admission in itself.

5   Throughout the globe-spanning and at times almost dream-like ride *The Crown* takes us on, and even as the cast switches completely every two seasons, one constant that stands out above all else is how much care was put into every single scene involving aviation. And there are a lot of them, to say the least. I have never seen a show so removed from transportation technology at its core, yet work so hard to make its transportation technology set-pieces so accurate. Each and every scene that an aircraft appears in just oozes passion for the topic.

6   Aviation was and still is a huge part of the Royals' lives. With the show kicking off in 1947 and ending its fourth season in 1990, it has aircraft, and the use of air transportation in general, seeded throughout. In a way, it gives a unique depiction of aviation history, including the progression from the piston engine age to the jet age, just as it does of UK history. While the writers definitely brandish their artistic licenses in crafting the show's plot, it doesn't do the same with the aircraft it features.

7   Not only was flying a staple act for Royals executing their duties, but there are large plot threads based around Prince Philip learning to fly and the death of many of his exiled family members aboard a Junkers Ju 52. Even a flashback of a flyover of Windsor Castle by Lancaster bombers during World War II is executed in a way in which you feel like you're standing right there looking up. Those types of scenes also make you happy you invested in good home theater audio as the sound editing is fantastic.

8   It is so refreshing to see reasonably accurate depictions of the real and somewhat obscure aircraft flown during each period of time the show covers, not just some convenient stand-in planes picked off the shelf. And they are all there. As you move through the series and the decades you will be showered with Vickers Viscounts, De Havilland Devons, and so much more. And yes, the RAF Red Arrows also makes an appearance.

9   All that being said, if you really dig through every aircraft and date, you will be able to find one or two discrepancies, but these would be extremely trivial in nature,

like an aircraft showing up a year before it entered service or in a different livery. A couple of flying sequences may be a tiny bit embellished, as well, but they serve as key plot points, so that is understandable. Overall, it's very impressive, and really, when it comes to production, this is what *The Crown* is all about – extreme attention to detail that immerses the viewer in a rich historic bubble. How they make this happen is fascinating in itself, but for the aircraft, they use a mix of real aircraft, practical sets, and CGI to make them come alive in the moment.

10 Just to underscore how much of a triumph *The Crown*'s homage to aviation is, take another big-budget show streaming right now that is actually all about aviation – *For All Mankind*. I am a fan of this show. It is engrossing and it doesn't play the alternate timeline/reality card too heavily considering it underpins its premise, which allows it to work well. The acting and characters are great. Yet, amazingly, what it is inconsistent at, at best, is actually depicting the aviation and spaceflight events the plot demands with believability and accuracy. It can go from engrossing to distractingly cheesy in an instant when it depicts flying sequences. This is not nitpicking from an aviation nerd, and it really takes you out of the moment with how cartoonish these scenes can get. I consider this a huge loss and overall missed opportunity for what is otherwise a good show. It almost seems that somehow *For All Mankind*'s production team thought that good drama could just paper over rickety aerospace action sequences. Clearly, I would beg to differ in this regard.

11 This comparison provides a stark contrast. That *The Crown* is not about aviation at all, but it goes so far and works so hard to give its aerospace-related moments such wonderful treatment, largely separates it from the pack. The fact that it does flying sequences remarkably better than a show about flight also proves that *The Crown*'s attention to detail is an equal opportunity proposition – nothing gets a pass.

12 There is so much in *The Crown* for other transportation fans, as well. The cars increasingly take center stage as the show moved forward – Jags, Aston Martins, Rolls-Royces, and Land Rovers galore. What's better is that they all "feel" lived in, replete with whining brakes and faded canvas tops. We also get great depictions of other forms of royal transportation, like the royal train and the iconic royal yacht, HMY Britannia.

13 So, if any of this interests you, you should start streaming *The Crown* immediately. While you may have come for the planes, trains or automobiles, you will end up falling in love with everything else this marvelous series has to offer.

New words

cinematic /ˌsɪnə'mætɪk◂/ *adj.* connected with movies and how they are made 电影的; 电影制作的

cast /kɑːst/ *n.* all the people who perform in a play, movie, etc. 演员阵容, 全体演员

premiere /'premɪə/ *v.* to give the first public performance of a movie, play, etc. (电影的) 首映; (戏剧的) 首演

dour /dʊə/ *adj.* not pleasant, interesting, or exciting 令人沮丧的; 无聊的

intrigue /'ɪntriːg/ *n.* the making of secret plans to harm someone or make them lose their position of power, or a plan of this kind 阴谋, 密谋; 诡计

melodrama /'melədrɑːmə/ *n.* a story or play in which very exciting or terrible things happen, and in which the characters and the emotions they show seem too strong to be real 情节剧

triumphantly /traɪ'ʌmfəntli/ *adv.* successfully 胜利地, 成功地

set-piece /'setpiːs/ *n.* part of a play, piece of music, etc. that follows a well-known formal pattern or style, and is often very impressive (戏剧、音乐等中) 有固定模式的精彩片段

ooze /uːz/ *v.* to show a lot of a particular quality or feeling 充满, 洋溢着

piston/'pɪstən/ *n.* a part of an engine that consists of a short cylinder that fits inside a tube and moves up and down or backwards and forwards to make other parts of the engine move 活塞

brandish /'brændɪʃ/ *v.* to wave sth., especially a weapon, in an aggressive or excited way 挑衅地或激动地挥舞 (尤指武器)

discrepancy /dɪ'skrepənsi/ *n.* a difference between two things that should be the same 差异, 不一致

livery /'lɪvəri/ *n.* the colors and designs that a company uses on its products or on vehicles that belong to the company 标志性颜色和图案

embellish /ɪm'belɪʃ/ *v.* to make sth. more beautiful by adding decorations to it 装饰, 修饰

engrossing /ɪn'grəʊsɪŋ/ *adj.* very interesting and needing all your attention 引人入胜的; 使人着迷的

underpin /ˌʌndə'pɪn/ *v.* to give strength or support to sth. and to help it succeed 巩固, 支持

nitpicking /'nɪtˌpɪkɪŋ/ *n.* the act of finding small mistakes in sb.'s work or paying too much attention to small details that are not important 吹毛求疵; 挑剔

rickety /'rɪkɪti/ *adj.* in bad condition and therefore weak and likely to break 摇摇晃晃的; 不结实的

galore /gə'lɔː/ *adj.* in large quantities 大量的, 很多的

canvas /'kænvəs/ *n.* strong cloth used for making sails, tents, shoes, bags, etc. 帆布

Phrases and expressions

cap off 结束	paper over 掩盖
in retrospect 回想，回顾	action sequence 动作场面
give sth. a shot 尝试做某事	in this regard 在这方面
to say the least 至少可以说	provide a stark contrast 形成鲜明对比
plot thread 情节线索	take center stage 占据中央舞台
home theater audio 家庭影院音响	

Proper nouns

Junkers Ju 52 容克-52运输机	Aston Martin 阿斯顿·马丁（汽车品牌）
Windsor Castle 温莎城堡	Rolls-Royce 劳斯莱斯（汽车品牌）
RAF Red Arrows 英国皇家空军 "红箭" 特技飞行表演队	Land Rover 路虎（汽车品牌）
	HMY Britannia (Her Majesty's Yacht
For All Mankind《为全人类》（剧名）	Britannia) 皇家游艇 "不列颠尼亚号"
Jag 捷豹（汽车品牌）	

Technical terms

piston engine 活塞式发动机	Lancaster bomber 兰开斯特轰炸机

Background information

Netflix 网飞/奈飞公司
Netflix, Inc. is an American entertainment services provider founded in 1997 by Reed Hastings and Marc Randolph. It offers a library of movies and television series through distribution deals as well as its own productions, known as Netflix Originals.

The Crown《王冠》（剧名）
The Crown is a historical drama web television series created and mainly written by Peter Morgan for Netflix. It is produced by Sony Pictures Television and Left Bank Pictures. *The Crown* portrays the life of Queen Elizabeth II from her wedding in 1947 to Philip, Duke of Edinburgh, until the present day.

CGI 计算机生成图像

CGI stands for computer-generated imagery, which is the use of computer graphics in art and media. These can be 2D or 3D animations, objects, or renderings; the type of art or media can be a movie, television program, video game, or simulation. CGI can be used in movies ranging from science fiction epics to quiet intimate dramas.

D ifficult sentences

1   **While the acting, writing, sets, and costumes all get much acclaim, one element of the show's world-building magic that doesn't get enough attention is its near-perfect attention to detail in how it portrays aviation's intersection with the Royals' lives. (Para. 1)**

Although so much of what you see and experience in this TV series is all highly praised, part of the show's extraordinary power is ignored in recreating a new world on screen. That is to say, the show almost perfectly depicts how aviation and Royals' lives are related, but this is overlooked.

2   **As it went on to win award after award I never found the time to come back to it – I wouldn't call myself a big enthusiast of the Royal Family and the never-ending aura of palace intrigue that surrounds it. (Para. 2)**

Although *The Crown* was given many outstanding honors, I didn't bother to watch it. I am not very interested in the Royal Family and the endless conspiracy living with it.

3   **What I quickly learned is that it not only was the show absolutely incredible in pretty much every way, but it really wasn't just about the Royal Family's notorious insider melodrama, and it was just as much a political drama as anything else. (Para. 3)**

I soon realized that the show was really not only fantastic in almost every aspect but also a political drama concerned with infamous stories about the Royal Family.

4   **While the earlier seasons of the show were definitely stuffed with period romance, the kind of which the last two seasons couldn't compete with due to the portrayed progression of time, the level of writing and the complexity of the storytelling evolved triumphantly throughout the series back half. (Para. 4)**

Although the first two seasons of the show were surely filled with historical appeal, which the last two seasons couldn't match because of the processing time, the writing techniques and the level of storytelling in the latter seasons were much improved.

5　**Throughout the globe-spanning and at times almost dream-like ride *The Crown* takes us on, and even as the cast switches completely every two seasons, one constant that stands out above all else is how much care was put into every single scene involving aviation. (Para. 5)**

The Crown takes us on a worldwide journey that is near-perfect fantasy. Though all the actors and actresses were completely replaced every other season, something that does not change is that much consideration has been given to every detail related to aviation.

6　**Aviation was and still is a huge part of the Royals' lives. With the show kicking off in 1947 and ending its fourth season in 1990, it has aircraft, and the use of air transportation in general, seeded throughout. (Para. 6)**

Aviation always plays an important role in the Royals' lives. The TV show starts its story in 1947 with the fourth season finishing in 1990. From beginning to end it features a variety of airplanes and air transportation.

7　**While the writers definitely brandish their artistic licenses in crafting the show's plot, it doesn't do the same with the aircraft it features. (Para. 6)**

The writers confidently make use of their imaginative skills when creating the storyline, but it is based on facts referring to aircraft.

8　**Even a flashback of a flyover of Windsor Castle by Lancaster bombers during World War II is executed in a way in which you feel like you're standing right there looking up. (Para. 7)**

There is a historical scene showing Lancaster bombers flying over Windsor Castle during World War II. It is done in such a vivid way that you feel you were actually present when it happened.

9　**As you move through the series and the decades you will be showered with Vickers Viscounts, De Havilland Devons, and so much more. (Para. 8)**

When you watch the series and travel through tens of years, you will see different historical planes like Vickers Viscounts, De Havilland Devons, and many others.

10 **All that being said, if you really dig through every aircraft and date, you will be able to find one or two discrepancies, but these would be extremely trivial in nature, like an aircraft showing up a year before it entered service or in a different livery. (Para. 9)**

After saying all that, examining the detail of every aircraft and its history, you can notice a few mistakes, but they are very minor. For example, an aircraft may appear a year earlier than it was actually available, or its colors and designs may be incorrect.

11 **Just to underscore how much of a triumph *The Crown*'s homage to aviation is, take another big-budget show streaming right now that is actually all about aviation – *For All Mankind*. (Para. 10)**

In order to emphasize just how successful it is for *The Crown* to show its respect for aviation, we can compare it to another TV series with heavy investment showing online, *For All Mankind*, which is focused on aviation.

12 **It almost seems that somehow *For All Mankind*'s production team thought that good drama could just paper over rickety aerospace action sequences. (Para. 10)**

For All Mankind's producers appeared to have believed that poor aircraft action scenes would be neglected if the story and acting were good.

13 **The fact that it does flying sequences remarkably better than a show about flight also proves that *The Crown*'s attention to detail is an equal opportunity proposition – nothing gets a pass. (Para. 11)**

The fact that *The Crown* deals with flying scenes obviously better than a show which is particularly focused on aviation shows that *The Crown* is very attentive to every detail, and nothing can slip through the net.

14 **What's better is that they all "feel" lived in, replete with whining brakes and faded canvas tops. (Para. 12)**

What is more impressive is that these cars appear to have long been used, including producing high-pitched noises when stopping, and being equipped with faded cloth roofs.

Exercises

Fill in the blanks with the words and expressions given below, and change the forms when necessary.

> immerse   acclaim   embellish   underpin   intrigue
>
> discrepancy   execute   depict   engrossing   triumphantly

1 Maritime transportation and air transportation are two essential activities that _____ global trade and mobility and are key to a sustainable socio-economic recovery.

2 The job of our hospital involves drawing up and _____ a plan of nursing care for older patients.

3 She failed to notice the _____ between the name on the check and the name on the driving license.

4 This beautiful palace, like every palace since the dawn of civilization, is often a place of _____ and calculation.

5 The exhibition features a total of 206 important ceramic items and oil paintings that _____ the stories of early trade and cultural exchanges between China and Europe.

6 She passes the test _____ as her mother predicts and, in return, is amply rewarded.

7 "Instead of using plastic accessories to _____ garments, we can start like growing onto our garments these new materials and more natural materials," Potts said.

8 While a few fans have witnessed the games in person in Olympic Games, the tight nature of competition has meant that these games have been some of the most _____ and engaging in recent history.

9 The purpose is to see that there is a need to _____ the mind and the body in order to be alleviated from the agonies of the present moment.

10 These artists and their defenders cited such popular _____ as proof of the aesthetic value of their works.

|| Translation

Part A   Translate the following paragraph into Chinese.

All that being said, if you really dig through every aircraft and date, you will be able to find one or two discrepancies, but these would be extremely trivial in nature, like an aircraft showing up a year before it entered service or in a different livery. A couple of flying sequences may be a tiny bit embellished, as well, but they serve as key plot points, so that is understandable. Overall, it's very impressive, and really, when it comes to production, this is what *The Crown* is all about – extreme attention to detail that immerses the viewer in a rich historic bubble.

Part B   Translate the following paragraph into English.

张明自2018年起致力于无人机拍摄。从商业视频到微型电影，他的工作均有涉及。无人机操作员，尤其自互联网广告和短视频繁荣发展以来，因创造性和舒适性而成为流行的职业。不过，无人机的拍摄并不容易，几分钟的视频背后是从设计到编辑的数天工作。小张喜欢他的"理想工作"，尤其是当朋友们称赞他的作品时。

||| Write a summary of the text in 120 words.

Directions: A summary should be written in your own words. It should contain only the ideas of the original text. Do not insert any of your own opinions, interpretations, deductions or comments into the summary.

|V Read the passage, select one word for each blank from a list of choices in the bank, and change the form when necessary. You may not use any of the words in the bank more than once.

> rumor   purchase   actor   monthly   license
> shuttle   qualify   play   aircraft   magazine

Tom Cruise got his private pilot's 1) _____ in 1994, and flying is a skill he began to acquire while 2) _____ Peter "Maverick" Mitchell in the Navy's Fighter Weapons School movie *Top Gun*. While in real life, Cruise would have been too small to 3) _____ to be a Navy pilot, he has bought several military 4) _____.

Most recently, he 5) _____ a World War II P-51 Mustang fighter plane and had "Kiss Me, Kate" written on the side. He told *People* 6) _____ that as a kid he carried a picture of a P-51 around with him. Guess dreams really do come true. Eco-lobbyists have nicknamed the 7) _____ "Emissions Impossible", as the Cruises have four planes now that they allegedly use like family minivans, 8) _____ kids around the world, going on 9) _____ shopping sprees, and popping in on the latest fashion shows. It's been said that Tom's even used the jet to pick up groceries, a(n) 10) _____ he hasn't denied.

V Critical thinking

Work in groups and discuss the following questions.

Aviation has found its applications in entertainment, among other things. For example, airships have been developed to meet tourist demand for sightseeing in the air. Compared with fixed-wing planes, airships have much lower costs. In addition, they can be free to fly almost everywhere, as they do not require a large airport. Therefore, they are a good choice for aerial tourism.

Suppose you are an airship designer, what facilities would you like to include in your design? In addition to tourism, what other applications would you like to take into account in your design?

⦿ Further reading

Airplane myths the movie industry made us believe

1 Who doesn't love a good movie? Nothing to take the edge off a stressful day like losing yourself in political intrigue, holding your breath during an intergalactic battle scene, or tearing up as long-lost lovers reunite. But as a pilot, nothing spoils my fun like a bad airplane movie myth. These frankly silly plot devices might make for an exciting movie, but have many travelers holding back in their seats. Time to set the record straight and debunk some myths.

2 **Airplanes are full of secret crawl spaces.** Worst offender of this urban myth: *Flightplan*, which features Jodie Foster on a crazy search through the secret airplane catacombs, making *Labyrinth* look a morning stroll to the bathroom. In reality, on-board real estate is at a premium. Every square centimeter is put to good use – leg room is saved for the passenger cabin. There are some larger aircraft that have access to the cargo hold, but only if you move some seats aside and roll back the carpet. And trust me – there's no hidden maze beneath your seats.

3 **Lose an engine and it's all over.** This one always gets my goat. Even single-engine aircraft can easily survive losing an engine; there are plenty of examples of planes losing their only engine and gliding in safely for a powerless landing. As for aircraft with two engines or more? Double as safe. Even if an engine fails during takeoff, pilots have a trained response to make it back to the ground in one piece – and, in fact, an airplane can fly safely with just the one engine.

4 **Turbulence? The cabin lights will flash dramatically, and the oxygen masks will drop!** My mom – bless her – clings onto the armrests for dear life at the first sign of disturbance. But although many people are afraid of it, turbulence is a natural consequence of flying. The Earth's atmosphere is rarely perfectly smooth, and

turbulence is simply the result of the many variations in airspeed, direction, and density. Think of it like bumps on the road. Airplane's wings are tested to withstand 150% of the maximum stress they will ever encounter – compared to that, turbulence is a piece of cake.

5 **A hole in the side of the plane will suck everyone out.** We can thank *Final Destination*, *Air Force One* and even James Bond for this particular fear. While it's true that the air pressure outside is lower than inside the aircraft, it's not enough to go full vacuum cleaner the moment you pop a hole in the side of the plane. After the initial rush of air (which will be about strong enough to blow around loose leaves of paper), the pressure will equalize and the air will just rush past the hole. Sure, it'll be cold, it'll be noisy and not all that pleasant – but no baddies will be sucked into oblivion while 007 is (literally) hanging from the rafters.

6 **Airplane doors can be opened in flight.** Maybe back in 1932, when aircraft weren't pressurized yet. Modern aircraft however are pressurized, and are designed to use this to their advantage. As the airplane climbs, the difference in pressure increases and the plug doors seal themselves shut. To open the door in flight, you'd need to be able to move 24,000 pounds of pressure – which comes down to about six cars or two fully grown elephants. Unless you have a team of strongmen at the ready, that door is not moving.

7 **The fuel tank is a ticking time bomb, waiting to blow.** As a pilot, *Con Air* was hilarious. The scene where the airplane barrels down the Las Vegas strip, losing its wings and leaving a fuel-infused inferno behind still makes me giggle. Airplane fuel is only flammable when it's sprayed; jet engines have complex nozzles and air swirlers in their combustion sections. The fuel is actually designed to be quite flame-resistant as a liquid – you could use it to douse a lit match. That waterfall of kerosene gushing from the airplane is more likely to extinguish a stray cigarette butt on the road than to cause an explosion.

8 **Lightning will blow up the plane.** Airplanes actually get struck by lightning more often than you'd think. On average, every airplane gets hit by lightning once a year; this comes down to more than 50 strikes a day, worldwide. And while it can be pretty

frightening for passengers, in reality it's not a big deal. Since the loss of a Pan Am Boeing 707 due to lightning in 1962, all aircraft have been designed to properly deal with it. There's no chance of it overloading the on-board electrical systems or creating a spark in the fuel tank. Airplanes are designed to conduct the electricity around the outside of the cabin and fuel tanks; they become part of the lightning bolt's route to the ground.

9 **Losing the autopilot will cause the aircraft to crash.** In *Cabin Pressure*, the flight's autopilot gets hacked, causing the crew to lose control over the airplane. Ridiculous, if funny. In reality, every aircraft system is designed with multiple levels of safety and the autopilot is no exception. Commercial passenger aircraft have two to three separate autopilot computers, each of which can fly the airplane on their own. And even if that fails – pilots are actually trained to fly a plane, not just control the autopilot. Think of the autopilot as a trained monkey that turns the crank on a mechanical organ. If the monkey passes out, the pilot can turn the crank themselves.

10 So that's Hollywood debunked for you. I hope I haven't ruined too many movie moments for you, but on the other hand, you might feel a lot more comfortable the next time you get stuck in turbulence. If you want to watch a good airplane movie, I would have to recommend one of my favorites, which despite its stupid gags is pretty accurate: the 1980 classic *Airplane!* Yes, I'm serious – and please don't call me Shirley.

New words

intergalactic /ˌɪntəɡəˈlæktɪk/ *adj.* between the large groups of stars in space 星系之间的

debunk /ˌdiːˈbʌŋk/ *v.* to show that an idea, belief, etc. is false 揭穿…的真相; 驳斥

turbulence /ˈtɜːbjələns/ *n.* irregular and violent movements of air that are caused by the wind 不稳定的强气流

bump /bʌmp/ *n.* a small raised area on a surface 隆起之处

oblivion /əˈblɪviən/ *n.* the state of being completely destroyed 完全被毁灭; 彻底被消灭

hilarious /hɪˈleəriəs/ *adj.* extremely funny 令人捧腹的; 极其滑稽的

inferno /ɪnˈfɜːnəʊ/ *n.* a very large and dangerous fire that is out of control 熊熊大火, 火海

gush /ɡʌʃ/ *v.* to flow or pour suddenly and quickly out of a hole in large amounts 喷涌, 喷出

crank /kræŋk/ *n.* a handle on a piece of equipment that you can turn in order to move sth. 曲轴, 曲柄

gag /ɡæɡ/ *n.* a performance by a comedian (喜剧演员的) 滑稽表演

Phrases and expressions

take the edge off 减轻; 弱化
tear up 差点哭出来
roll back 卷起
get sb's goat 使某人十分恼怒
in one piece 安然无恙
cling onto 紧紧抓住
be sucked into 卷入; 吸入

come down to 归结为
at the ready 随时可用
barrel down 高速移动
get struck by 被击中
be no exception 不例外
pass out 晕倒, 失去知觉

Proper nouns

Flightplan《空中危机》（电影名）
Labyrinth《魔幻迷宫》（电影名）
Final Destination《死神来了》（电影名）
Air Force One《空军一号》（电影名）

Con Air《空中监狱》（电影名）
Pan Am 美国泛美航空公司
Cabin Pressure《紧急迫降》（电影名）
Airplane!《空前绝后满天飞》（电影名）

Technical terms

passenger cabin 客舱
powerless landing 无动力着陆
plug door 塞拉门

fuel tank 燃油箱
air swirler 空气旋流器

Exercises

| For each of the following unfinished statements or questions, choose the most appropriate answer from A, B, C, or D according to the text.

1 According to the text, what makes the author feel extremely annoyed?

   A. Losing an engine and all being over.

   B. Flying a single-engine plane.

   C. Gliding a plane without an engine for a safe landing.

   D. Shutting down the engines of a plane during takeoff.

2 The fuel from the airplane is _____.

   A. less likely to be lit when it is sprayed

   B. more likely to put out a lit cigarette

   C. kind of fire-resistant when it is preserved in the tank

   D. quite impossibly lit in any case

3   The number of the airplanes hit by lightning per year is around _____ worldwide.

A. 50

B. 1

C. 365

D. 18,250

4   What are designed to fly the aircraft safely if pilots are unavailable on the commercial passenger aircraft?

A. Autopilot computers.

B. Cranks.

C. Trained monkeys.

D. Control towers.

5   Theoretically speaking, pilots are trained to _____ if the autopilot computers are out of work.

A. fix an autopilot computer

B. control a monkey to fly a plane

C. set up an auxiliary autopilot computer

D. fly a plane themselves

‖ Questions for discussion

1   How can passengers get into the cargo hold in some larger airplanes?

2   Why shouldn't passengers be so anxious about turbulence?

3   Will a hole in the side of the plane suck everyone out? Why or why not?

4   What will happen if someone wants to open the door of a plane in flying?

5   What has been done to make the fuel tank safe?

⊙ Practical writing

✎ How to write meeting minutes

Meetings are always necessary at any organization that is serious in performing well in its field. Hence, the meeting must be recorded in the form of minutes for the relevant parties to refer to and be reminded of what has transpired in the meeting and the necessary actions to be taken as a follow-up.

Meeting minutes are usually formally written and distributed to the attendees for further action after the meeting. This task is usually performed by a secretary who sits in the meeting, recording down all that is discussed and decided by the meeting attendees. The meeting minutes may also be distributed to the absentees of the meeting. The template of meeting minutes may contain the following: date and venue of the meeting, purpose of the meeting, attendees/absentees, start and end time of the meeting, issues, time allocated for discussion, important points of discussion, and conclusions.

The meeting minutes can be long and detailed depending on the topic of discussion during the meeting, and the next meeting date can be included in the meeting minutes.

Exercise

Now suppose you are a secretary of a research institute, write the meeting minutes based on the following information.

2021年9月23日下午4点在第一会议室开会，会议由Mr. Falk主持，参会人员包括Mr. Baum、Ms. Dlugatz、Mr. Fenster、Ms. Liu和Ms. Sun，Ms. Penn缺席。在会上秘书宣读了上一次会议记录（8月24日），全体参会人员讨论决定：由Ms. Dlugatz而不是Ms. Penn对实验室数据做调查，由Mr. Fenster对实验室所做的调查进行总结并写一份报告递交研究所。下次会议时间定为10月22日下午3点15分，地点在第二会议室。